G. Forster
F C I O B Chartered Builder

Building organisations and procedures

Second Edition

Pearson Education Limited,
Edinburgh Gate, Harlow,
Essex CM20 2JE, England
and Associated Companies throughout the world

First published 1978
Second edition 1986
Sixth impression 1994
Seventh impression 1995
Eight impression 1998
Ninth impression 1999

British Library Cataloguing in Publication Data

Forster, G.
Building organisation and procedures. – 2nd ed.
1. Construction Industry – Great Britain
I. Title
338.4'7624'0941 HD9715.G72

ISBN 0-582-41373-7

Printed in Malaysia (VVP)

Contents

Chapter 4 Design principles and procedures

Chapter 5 The town planning situation

Chapter 6 The building act/regulations

Chapter 7 Safety, health and welfare on site

Chapter 8 Measurement procedure

Chapter 9 Contractor's pre-tender work

Chapter 10 Contractor's pre-contract work

Chapter 11 Contract work and other considerations

Acknowledgements

We are grateful to the following for permission to reproduce copyright material: the Royal Institute of British Architects Publications Limited for material adapted from *Plan of Work for Design Team Operations*; the Royal Institution of Chartered Surveyors; The Building Employers' Confederation; Her Majesty's Stationery Office for materials from Acts and Statutory Instruments, and data from issues of *Housing and Construction Statistics* and *Annual Reports of H.M. Chief Inspector of Factories;* Wickens Construction Group Ltd, Chertsey, Surrey.

The author would also like to thank his wife Shirley for typing the manuscript, and is indebted to Michael Gardner, MCIOB, for supplying details for the section dealing with Measurement Principles.

Chapter 1

Structure of the building and civil engineering industry

1.1 Types of work undertaken by construction firms

Construction firms either operate as building or civil engineering concerns, or both.

Limitations are placed on many construction firms because of their size, small firms being unable to tender (bid) for larger contracts because of their limited resources regarding capital and experience. On the other hand, large firms prefer to be involved primarily with major contracts as they are not normally geared to deal with small works of less than £1 m. The larger companies, however, are sometimes forced by circumstances to tender for any type of work outside their normal range during a recession on the home market, or seek more work abroad.

With the different legislation which is forever being introduced, particularly the Employment Protection Act 1975, the Employment Protection (Consolidation) Act 1978 and the Employment Protection (Amendment) Act 1979, also the Employment Act 1980 and 1982, work must be plentiful in order to retain a company's carefully built up workforce, otherwise the consequences when redundancies occur is an added unnecessary expense to a contractor. Also, when work becomes plentiful the contractor needs to be able to respond immediately when new contracts are won by having the correct number of key employees available at a moment's notice. It is obvious that too many firms now rely heavily on labour-only subcontractors to limit the need to worry about redundancy and employment protection, as the subcontractors can be dismissed at the end of a contract without legal or union repercussions, which it is believed by many, is an important advantage to the main contractor.

The work which is undertaken by contractors for clients is usually divided into two sectors, which are:

1. Private sector.
2. Public sector.

The contract work carried out in the private sector is for private individuals, sole traders, firms of partners, companies and corporations. In the public sector the clients can be the National Government, Local Authorities and other public bodies – British Rail, British Telecom, CEGB, Gas, Water, the National Coal Board and many others.

Numerous contracts for local authorities are undertaken by their own direct labour departments, the contracts usually awarded being in the field of new house construction, and maintenance of their existing houses. This practice is repeatedly condemned by the Building Employers Confederation (BEC) because it naturally partially deprives them of their 'bread and butter'. It is argued, however, that while some authorities' direct labour departments are unsuccessful in doing the work economically others are succeeding in operating as viable concerns.

The value of output for new public dwelling in 1983 was £1129 m. Contractors undertook about £1000 m. of this work, leaving the remaining £100 m. to direct labour departments of the various local authorities.

Housing, in both the public and private sectors, accounts for approximately one-third of the construction output of the United Kingdom. The value combines the building of new and the maintenance of old houses. See Fig. 1.4.1.

With the building of huge housing estates particularly for local authorities, including New Town Development Corporations, the whole spectrum of construction work is brought into being; for example: roadworks, sewers, street lighting and other services. Also, the types of buildings and structures required for the intending inhabitants increases depending on the density and size of the development. Shops of all types may be needed and can be situated with access from pedestrian precincts, along with community centres, churches, various schools, medical centres, libraries, public houses, sports areas or centres and play areas. Light industrial complexes may then be situated within easy access from the housing estates to provide local work.

System building has played a major role in the output of the industry and local authorities have tended to use industrialised components in their building programmes. The systems have been best used for dwellings, such as: detached, semi-detached and terraced houses; flats built on the high- or low-rise principles; schools; community centres and office blocks. The Easiform, Skarne, and Jesperson systems, to mention but a few, are systems which have been used extensively in the past.

Timber-framed housing is now a more popular type of construction process and has developed rapidly since about 1975. There have been problems with the earlier timber-framed houses, but better design and workmanship, it is assumed, have solved many of the problems.

To help offset the costs in their development programmes, numerous local authorities have formed consortia, particularly when it was necessary to build schools, colleges and community centres. This meant that through their collaboration they exchanged ideas and information and also arranged to choose suppliers for building materials and components. The whole design problems were also discussed and modified in the light of each authority's experiences. These collaborations still exist along with the systems designed, and they operate under such names as: the Consortium of Local Authorities Special Programme (CLASP), the Second Consortium of Local Authorities (SCOLA), the Metropolitan Architectural Consortium for Education (MACE), the South-East Architects Collaboration (SEAC), the Consortium of Method Building (CMB), the Organisation of North-West Authorities for Nationalised Design (ONWAND), the Anglian Standing Conference (ASC) and the Consortium of Local Authorities Wales (CLAW).

It can be easily appreciated that if a consortium of local authorities designs one basic structure, and all have access to it, cost savings can be considerable.

The centres of many towns are controlled by the local authorities and the land is leased to the various traders in the high streets. When the lease expires certain areas are allowed to run down and slum clearance and demolition work forms a considerable value of a contractor's output. Authorities usually have plans for new development leading to the replacement of the old Victorian shops, fire and police stations, swimming baths and schools. These, invariably, are replaced by constructing ultra-modern town or city centres with shopping precincts, supermarkets, hotels, theatres, sports centres, bus stations, underground and multi-storey car parks, offices and other commercial buildings. Each authority appears to attempt to out-develop the other, or at least tries not to be left too far behind. Prestige buildings and development are encouraged which would add to the influx of visitors, thereby bringing business into the community. Such buildings are universities and colleges, hypermarkets, exhibition halls such as the National Exhibition Centre in Birmingham, cathedrals (similar to that most recently built in Liverpool), and marinas like Brighton's. In 1985 a new Stock Exchange in London was completed costing £157 m.

Adjoining buildings to development areas, while structurally sound, are sometimes modernised by the lessee (tenant) or owner by refurbishing the building, which is to say, to completely up-date the structure regarding in particular the doors, windows, and the full range of interior fittings and services. The complete interior layout may even be altered with a wall being removed here and a floor strengthened there. A refurbishing contract can amount to millions of pounds worth of work for one building alone in the heart of London. Old structures are preserved and the surrounding areas are developed in an attempt to maintain our historical heritage and as an attraction to tourists, such as: Thomas Telford's old London Dock – now a marina; the old Covent Garden – now a small business and entertainment area, etc.

Warehouses, factories, breweries, distilleries, prisons, garages and hospitals for the public and private sectors adds to the builders' and civil engineering contractors' workload. There are, however, other more spectacular contracts which swell the turnover of contractors' businesses, namely: power stations (Dinorwig), airports, and military establishments and installations. This work usually requires the exceptionally organised contracting firms' services. Bridges such as the Humber one costing £500 m., and projects such as the Thames Barrier, which cost £400 m., are other types of work.

More recently, due to Britain's sea-bed gas and oil exploration, a new technology has arisen – that of the design, construction and anchoring of off-shore drilling platforms, oil storage chambers and associated work. The platform construction yards in themselves, while being constructed, are major civil engineering projects. These are new industries to this country, and the techniques used are still in a state of infancy throughout the world. It requires very experienced specialists and contractors to undertake the work. Some of the platforms are made of concrete built on the caisson principle, but the great majority are of steel. Computers are used to successfully predict the weather to be expected for the final transportation of the platforms to their destination in the North Sea, and they may be eventually positioned accurately using Space Reliability techniques which are used by the space flight industry. When the platforms are finally allowed to settle on to the sea-bed, work begins immediately on the anchoring of their legs to piles which have to be driven deep enough to counteract the rough seas expected in the North Sea.

Other special work which can be expected by contractors, after the recent serious droughts, is for the construction of more reservoirs, and the possible formation of a national water grid (or at least a semi-national grid) similar to that which operates for electricity and gas. Nuclear waste bunkers must figure in construction work in the future, as must on-shore/off-shore windmills for generating electricity. Finally, it is expected that a new Severn bridge will be required due to serious structural faults which may make the existing bridge a limited means for crossing the river.

1.2 Financing the industry

If sole trading contractors require to finance work in the construction industry, the first sources (1–5) which are automatically available are:

1. *Personal capital*

One's own capital in the form of personal savings or fixed assets (premises, vehicles or plant).

2. *Profits*

Profits from past contracts.

3. Arrears

The use of heating, lighting and telephone systems which requires payment in arrears; therefore, credit facilities are afforded by the public services undertakings.

4. Employee advance work

Work in advance of payment by own employees.

5. Advance work – outside labour

Work in advance of payment by subcontractors, nominated subcontractors, labour-only subcontractors.

It is of course wrong if an employer expects his workforce to labour all week or month if he knows that at the end of these periods there are insufficient funds to pay the wages or salaries. Employees normally, however, have a reasonable amount of protection in law, and are safeguarded against loss of earnings due to the employer being unable to meet the demands of the payroll costs. Wages and salaries are one of the first payments made in the event of the employer going into liquidation. The assets of the business are sold off to help pay the debts to the creditors – the employees who are owed wages or salaries being in this category.

Further methods (6–7) of raising and using finance requires a degree of effort and personal tact on the part of the entrepreneur (sole trader or risk taker). Approaches could be made to various bodies to obtain credit facilities. These can be outlined under the following:

6. Trade accounts

Agreements to open trade accounts with merchants and manufacturers, which allows the contractor to purchase goods at any time without paying for them until the 'statement' is received at the end of the month.

7. Plant hire

Plant-hire firms could give a similar service, i.e. the use of plant and vehicles on credit terms, the account being settled at the month's end.

In both of these methods trade and cash discounts could be arranged.

(*a*) *Trade discounts:* These are usually offered to those in business, the amount varying from 5 to 40 per cent, depending on the type of business and the quantity purchased or used.

(*b*) *Cash discounts:* These are offered as an inducement to pay debts promptly, the amount being in the region of 2–5 per cent. If the contractor is careful to pay his debts as soon as the 'statement' for the goods is received, the cash discount savings could be put to further use within the business. This is an additional financial source.

When capital expenditure is necessary to increase the fixed assets of the business, i.e. premises, vehicles, plant and machinery, and furniture and fittings, there are a number of routes open to explore the possibilities for borrowing. These are (8–12) headed under: Building societies (mortgage); Banks (loans and overdrafts); Finance houses (loan); Insurance companies (mortgages, etc.); Friends; Relations.

8. Building societies

Mortgages for the purchasing of buildings and land can be negotiated through these organisations. They will also extend their facilities to include alteration and extension work to existing properties, and to the development of land. The repayment is usually on the best terms possible compared to all the other long-term methods. There is, however, a drawback, inasmuch as there is usually a period of between six to twelve weeks' delay while the mortgage agreement is drawn up by the Society. Also, the availability of a mortgage depends on the degree of risk the Society thinks exists when it is contemplating a businessman's request for a mortgage. Generous investment terms, such as high interest rates, will entice investors to increase investments within the societies, therefore making more money available for mortgages.

9. Commercial banks

There are many facilities which these banks offer and they are shown in Fig. 1.2.1. As a financial source they not only exist to arrange mortgages, but more usually they will provide Loans or Overdrafts for customers who wish to borrow money, the differences being outlined below:

Loans: A customer may arrange between himself and the bank manager to borrow a lump sum of money which is either handed over the counter, or can be transferred from the Bank's funds to his current account, to be used at will. A charge of between 10–15 per cent interest is made by the bank on the total sum borrowed, until it is completely repaid. Another method would be to arrange stage payments and the balance still outstanding would have interest charged to it. Some form of security, such as a life insurance policy, or deed of ownership of property may be required by the bank.

Overdraft: Permission for this form of borrowing must also be given by the bank manager, and similar percentage interests and collateral would apply. Although in this system a lump sum is not transferred to the client's account, a fixed amount of money may be overdrawn from the current account, and interest is charged usually on a monthly basis. If the account is not overdrawn in any month then no interest is charged.

Other banking systems

These are Trustees Savings Banks and the Post Office. The former was primarily a savings system but now has similar facilities to a clearing bank. The latter system has a limited use for saving money but can be used for paying debts

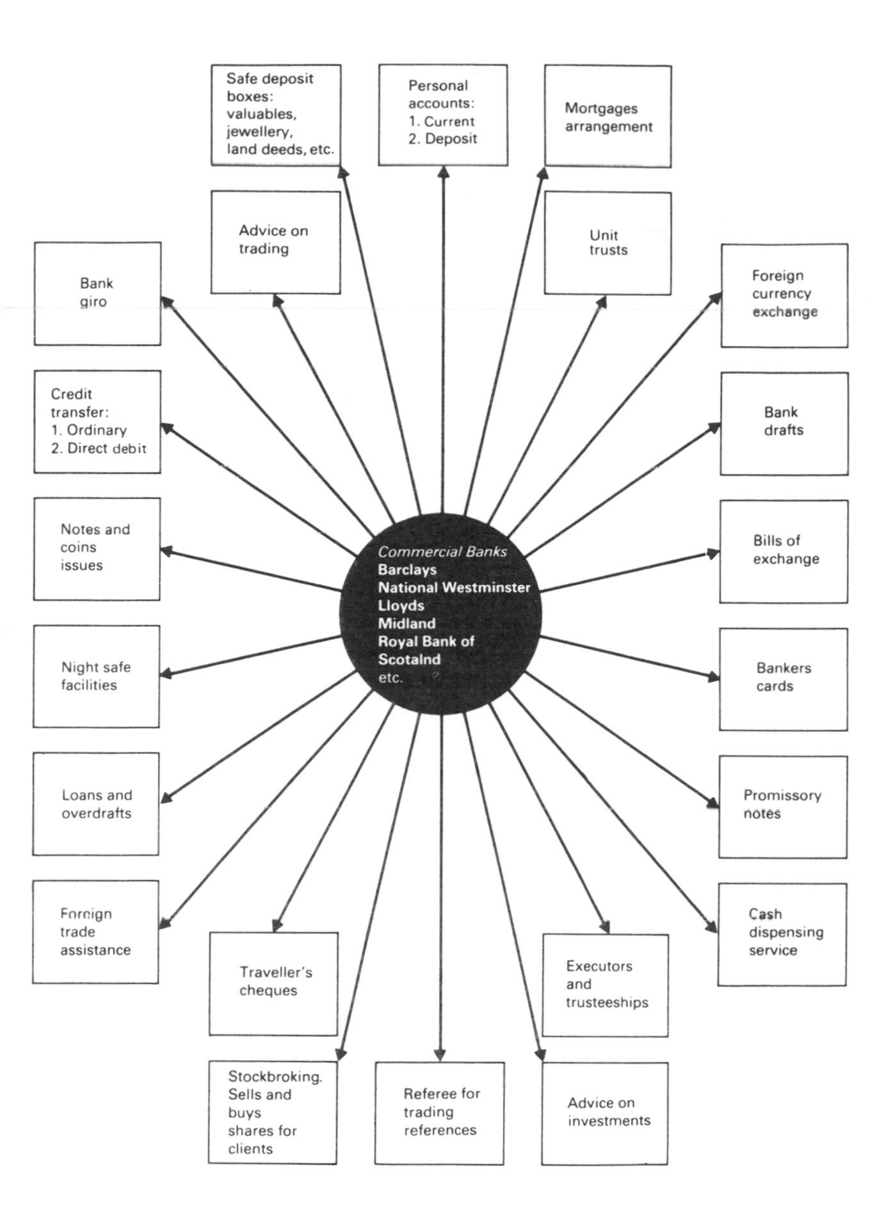

Fig. 1.2.1 Facilities offered by commercial banks
Note: The prime function of a commercial bank is to make profits for its shareholders

through money orders, postal orders, telegraphic money orders and the Giro system. Postal orders can be made out for small sums up to £10. Money orders on the other hand can be used for payments of larger amounts. Also, money orders can be telegraphed to speed up payment.

With the Giro system an account must be opened by the customer (businessman) and is used similarly to a commercial bank for the payment of debts, or the receipt of cash from customers through their bank or post office account, either by using transfer forms or the direct debit transfer system. Loans cannot be negotiated through either of these systems.

10. Finance houses or companies

Many firms and private individuals use the facilities offered by these organisations for the purchasing of vehicles and plant. However, they are used for other hire-purchase transactions.

The merchant or trader can arrange loans through the finance houses for most clients or customers. The way in which this system works is that if a vehicle is to be purchased by a customer who unfortunately has insufficient cash to cover the cost, but is willing to give assurances that repayments will be made with weekly or monthly sums of money, until the total cash borrowed is written off, a standard 'agreement form' would then be signed by the customer, and this agreement would be between him and the finance company, and not between himself and the merchant or trader.

Once the agreement is completed the merchant/trader receives the money due to him on behalf of the client from the finance company. The finance company is then repaid through the hire purchase agreement method, including interest which is normally much higher than borrowing from a bank.

It is interesting to note that a deposit is normally required from the purchaser, who in fact would be known as the hirer, until such time as the goods (commodities) are paid for fully.

11. Insurance companies

Mortgages for land and premises can be arranged but most usually on an endowment policy system. This tends to be more expensive than by an ordinary mortgage scheme through a building society. It does of course give insurance cover on the value of the mortgage in the event of the insured's (mortgagee) death. This means that dependants of the insured would not have further amounts to pay for the outstanding mortgage on the land or premises.

12. Friends and relations

These sources of finance are obvious. The point to make here is that a proper system of repayment should be made, with interest, or at least a share in the fortunes of the business should be given commensurate to the capital invested by these people.

There are other sources available which cannot be placed in the same category as those previously described. Some of them are additional ways of borrowing money or using capital, during the day-to-day running of the business, others are by utilising one's own resources and transferring fixed assets into cash, for use within the firm to keep production flowing. These are:

13. Leasing

Instead of tying up valuable cash in the purchasing of one's own business premises, leasing or renting may be the answer, provided that suitable accommodation can be found which is of the correct size, position and rental value. On the other hand, property such as premises is a good investment, because its value keeps pace with inflation, and is an asset which can be used as collateral in the event of the need to borrow money.

14. Lease-Back

If a business has a quantity of fixed assets they can be sold at any time to help pay off debts or to raise liquid capital. If the assets which are sold reduces the efficiency of the business, then hiring similar items as and when required may be the answer. Another technique would be to sell one's own business premises to a finance company, with the agreement that the premises are retained for use indefinitely or for a fixed period by the firm, to conduct its business. Naturally, this requires a 'lease agreement' on the property, and therefore a rent would be paid weekly or monthly until the duration of the lease runs out, or is terminated by agreement on both sides. An agreement could be made to buy back the property at a later date at the market price, or a fixed percentage increase, whichever is the greater.

15. Leasing-out

Some firms hire out their plant and equipment to subsidise their own income from contract profits. Patent rights or copyrights can be issued on licence to those who wish to use particular ideas or techniques.

16. Invoice discounting

Factoring firms or finance houses will in some circumstances purchase the invoices from a business immediately they are prepared and are to be sent to debtors, but at a reduced rate (discounted). Factoring firms then take on the responsibility of collecting the money owed on the invoices.

With the adoption of the invoice by the factoring firms, cash is made available promptly so that a business has immediate use of cash to further its aims, and does not have the responsibility of waiting to collect the debts at the end of the month. Any bad debts which accrue from the invoices are borne by the discounting firm.

In the case of a debtor taking too long to pay what he owes, a debt collector – sometimes nominated through the local Chamber of Commerce – could be brought in to pressurise for payment, reaping about 10 per cent for any dues they manage to extract from slow payers.

15. CITB

The Construction Industry Training Board assists firms to train operatives and other employers by giving grants for this purpose. This method offsets the costs of training personnel for the contractor. Some of the funding for young people comes from the Manpower Services Commission (MSC).

16. Consortium

This is resorted to when the resources of firms are inadequate to undertake work of a specialist or very involved nature. Two or more firms or organisations combine their resources for a fixed period to work on a project which is beyond the scope of either firm. Smaller firms may even join forces to 'bid' for work against the larger contractors.

A contractor whose main efforts are in environment services, may combine with a civil engineering firm in bidding for work on the development of a shopping precinct. On the other hand, a general builder may join forces with a large plant-hire firm to undertake a civil engineering project – the profits being shared depending on the proportion of capital and the expertise used by each member of the consortium.

The advantage of forming a consortium can be shown that through limited resources, the same can be put to maximum usage, without incurring huge expenses in gearing the firms separately to do the work.

17. Government grants

These are made under certain circumstances, but especially for providing 'performance bonds' for contracting firms obtaining work abroad. Large contractors are called upon to show good faith by placing large cash guarantees (bonds) before these overseas clients. The bonds are repaid on the satisfactory completion of the contract.

The British Government helps in every way it can to boost exports and most firms take advantage of the facilities which are made available.

18. Stage or interim payments

A contractor can agree to be paid at the following times:

(*a*) When work is completely finished.
(*b*) At various stages of the work, such as substructure completed, superstructure completed and then possibly internal finishes.
(*c*) Monthly, known as interim payments.

The value of the work can be agreed with the client before the first two methods of payments are made. With the third method the client's professional quantity surveyor agrees a value of work done at the end of the month. When the work is valued an interim certificate is signed to show the work certified, and a copy of the certificate is sent to the architect for approval; the client then becomes liable for payment for the value shown thereon.

When the money is received (within 14 days) from the client, it can be used to pay amounts owing to the manufacturers, suppliers, subcontractors, and employees, in addition to further financing the next stage of the work. It is normal for the main contractor to deduct up to 5 per cent from the money owing to the subcontractors for attendance and special attendance given to them.

19. Partnerships or companies

The formation of either of these firms from a sole trading concern would usually lead to the injection of additional capital and resources into the business.

It must always be remembered that the entrepreneur should not pay himself/herself an inflated salary, particularly when the firm is not progressing as was planned. Also, withdrawals from the business bank account for one's personal use must be minimised. Lastly, part, if not all, of the profits should be ploughed back into the firm if it is to continue as a viable business.

1.3 Exports and contracts abroad

British construction firms have enjoyed a reasonable measure of success abroad in the past fifty years, but in comparison to their almost monopolistic successes prior to this period, particularly in countries of the old British Empire, then it can be said they have lost ground to the more competitive firms of other countries. No one thought it more necessary than during the period of recession in the industry on the home market of 1975–6 to win work abroad in order to survive. Many firms, especially the large ones, realised years ago the importance of establishing bases abroad even though they were maintained, in many cases, on a small scale.

During the economic booms in Britain the major building and civil engineering companies, and even the firms specialising in design services, continued cautiously trading in other parts of the world while simultaneously trading on the home market. This can be termed planning for the future, and has since been proven justified. Operations abroad have been stepped up to utilise fully the resources of the companies or firms when the previously described recession occurred. The experiences so cautiously gained in the past years while dealing with foreign governments on a small scale have given the national firms of Britain a 'head and shoulders' start over other less organised rivals from home and abroad. Some rival companies have since paid the price for not diversifying their interests enough, and have either gone into liquidation, or have been voluntarily taken over by the better organised concerns.

Each foreign government has its own peculiarities of doing business and the firms who have learnt how to deal with them, by experience, are now making substantial strides by winning huge multi-million pound contracts for all types of building and civil engineering work. This is a form of export and Britain must be successful at it to survive. It is due to the larger firms' foresight that now they find themselves in a lucrative market, particularly inside the Organisation of Petroleum Exporting Countries (OPEC), such as Iran, Saudi Arabia, Federation of Arab Emirates, Bahrain, Nigeria, Libya, Algeria, Kuwait, Qatar, Dubai, Abu Dhabi, Oman, Iraq, Indonesia and Venezuela.

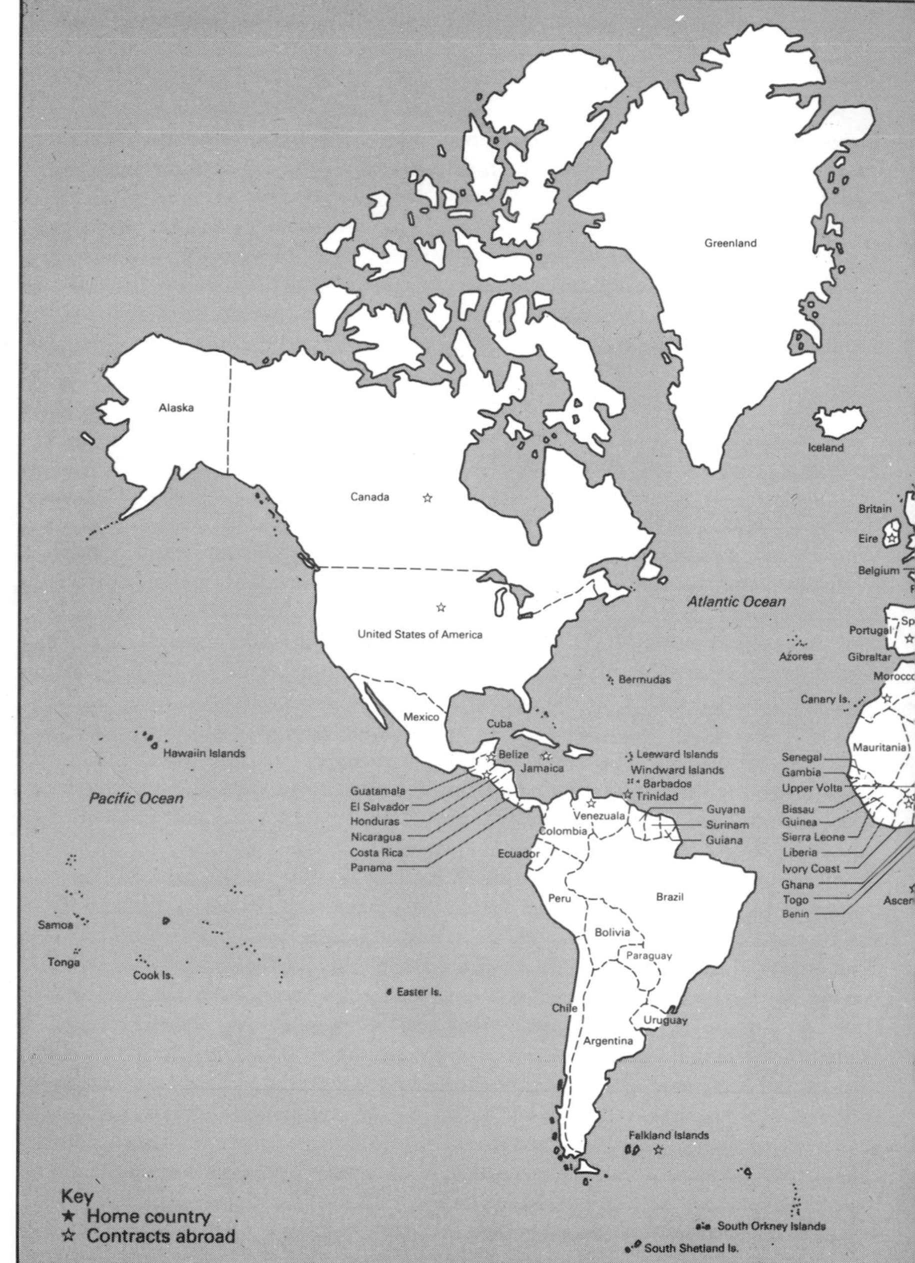

Fig. 1.3.1 Map of the world showing contracts abroad, 1983

Finland
Sweden
Baltic Sea
Holland
Germany
Austria
Union of Soviet Socialist Republics
Mongolia
China
Japan
Syria
Lebanon
Israel
Jordan
Turkey
Greece
Cyprus
Iran
Iraq
Kuwait
Bahrein
Qatar
Afghanistan
Pakistan
Nepal
India
Egypt
Saudi Arabia
Oman
South Yemen
Yemen
Federation of Arab Emirates
Burma
Hong Kong
Vietnam
Thailand
Cambodia
Malaya
Singapore
Sumatra
Borneo
New Guinea
Java
Indonesia
Pacific Ocean
Indian Ocean
Chad
Sudan
Central African Republic
Uganda
Ethiopia
Somalia
Kenya
Zaire
Tanzania
Seychelles
Angola
Malawi
Zambia
Mozambique
Zimbabwe
Botswana
Madagascar
Mauritius
South Africa
Australia
Gilbert Is.
Fiji
New Hebrides
New Zealand
Auckland Is.
Southern Ocean

Some contractors undertake the contract work abroad on a 'Management Fee' or Contract Management basis. They provide the know-how, management skills and management personnel, but arrange for the provision of plant and materials with other exporters, and use local labour and materials where possible. Some of the contracts are too large or complex for any individual firm to be able or willing to place a bid for it, because of the inherent risks, such as political instability, delayed interim payments, management problems regarding communication difficulties, language barriers, and being subjected to the law of foreign lands. The larger firms either undertake work as a 'Joint Venture' (Consortium) or a British firm may agree to combine with a firm of the country in which they wish to work. Profits are naturally shared out depending on how much responsibility and financial backing each firm has committed to the project. Figure 1.3.1 shows the countries in which British firms worked in 1983.

French, Dutch, West German, Italian, South Korean and American construction and design firms are the biggest competitors for building and civil engineering contracts abroad. Because of past work records, however, and in particular their competitive prices, British firms still maintain a reasonable measure of success in obtaining work.

Financial backing is necessary for firms operating abroad and this can be obtained, apart from their own resources, from commercial and merchant banks. The Government's Export Credit Guarantee Department will not only guarantee loans from bankers to exporting companies, but will repay the debts owed by foreign clients on their default to an exporter.

As an encouragement to firms contemplating seeking work in other countries the Construction Exports Advisory Board is available to give advice on the best approach to be made, and the countries to be avoided.

At the present time it is the OPEC countries – sometimes referred to as the third world countries because of their rapid rise to prominence and riches, when they artificially made the fuel they produced scarce and therefore expensive by restricting output – that are rapidly undergoing changes never before seen anywhere in the world. Their new-found wealth is mainly being used in huge building and development programmes of which the industrial countries' contractors are benefiting.

Unfortunately, due to the overproduction of oil in the early 1980s and the inevitable reduction in the price of oil, some OPEC countries have had serious reductions in their oil revenue which has seriously affected their development programmes, and British companies have lost lucrative contracts or have had difficulty in extracting payment for work done.

The earnings by construction companies abroad in 1974 was £459 m. In 1980 it was £997 m. and in 1983 it was £1018 m. The consulting engineers of Britain are also contributing to the nation's exports. The value of their earnings amounted to £189 m. in 1980. Even British town planning consultants are being called upon to advise on the layout of new town development in Arab and other developing countries which adds to the wealth of this nation.

The construction industry is the barometer of the country's prosperity, and success at home, but particularly abroad, is necessary to encourage every one of us.

1.4 Statistics on the industry

The amount of work undertaken by construction firms depends on the volume of finance which is available both in the public and private sectors. The dependability of the finance relies to a great extent on how well the country as a whole is doing regarding its Balance of Payments. The more it exports in comparison to its imports, the more the country prospers and usually this leads to an increase in spending on buildings and other structures.

The 'stop-go' attitudes of alternating governments seriously affects the work of the various construction firms. One year more is being spent on hospitals, the next year more money is made available for roadworks. Many firms find it absolutely necessary to diversify their operations so that they can meet the changes which occur frequently due to the successive governments' attitudes.

During an economic boom capital is poured into industries generally, the outcome being that the building and civil engineering contractors find their workload plentiful. It is a recognised fact that when there is a large increase in spending on big Government or local authority projects a country is prospering. It is therefore safe to assume that the construction industry is the barometer of the country's fortunes. When British industries and commerce fail to win exports, and the balance of payments tilt to a worsening position, there is inevitably a 'tightening of belts' to tide a firm over until the position again reverses in its favour. Immediately it is the building programme which suffers and the result is a reduction of contract work which is the life blood of the building and civil engineering contractors. With little work for which to tender the contractors may have to cut back their workforce which inevitably leads to redundancies. The small and medium-sized firms which are usually built on the family business traditions and have survived for many decades without venturing to expand, are usually the last to feel the pinch if there is a reduction in the availability of work. Also, the larger firms who vary the type of work they undertake can meet the ever changing needs due to the constantly changing attitudes of the Government. It is the speculation firms which grow during a period where housing demands are high that crash first. Naturally, mergers and take-overs have also contributed to the reduction in the number of construction firms from 84 000 in 1965 (one-man businesses to businesses which employed over 1000 employees) to 73 500 in 1970. There was then a dramatic rise to 96 000 in 1973, and the number will have dropped to the 70 000 region by the year ending 1976. Because of the increase in self-employment within the construction industry, records for numbers of businesses are now incomplete in 1984.

Statistics are produced by the Government in its *Annual Abstract of Statistics* and the *Housing and Construction Statistics* (quarterly) publications which are available from Her Majesty's Stationery Office. The figures given in the publication show the trends of the output in the different spheres of construction work. The figures could be used to forecast where best to deploy a firm's energies.

Statistical charts are best prepared to give a clearer pictorial representation of the many figures given in the various publications. Numbers in themselves are poor illustrators, and are difficult to assess and digest to show what is happening in a certain field of activity. Scarcity of a commodity, such as building materials, can also be highlighted so that these shortages can be planned for when a new contract is to be undertaken. A shortage of bricks similar to that which occurred in 1972 would delay the work of any contractor if orders were not placed sufficiently ahead of requirements.

The charts which are best used, examples of which are included later, to give a satisfactory pictorial view of what is developing can be chosen from the following:

(*a*) Graphs;
(*b*) Pie charts;
(*c*) Combined bar charts;
(*d*) Bar charts;
(*e*) Histograms;
(*f*) Pictographs.

Output within the construction industry can be affected by other criteria. Adverse weather not only affects contractors on-site but also the producers of raw materials. The brick manufacturers' output is restricted in extremely wet weather, and concrete producers would find it necessary to reduce output because of the smaller demand in extremely frosty weather. Another factor controlling efficiency on-site is labour relations (Industrial Relations). When disputes occur, output drops. Disgruntlement sometimes leads to walk-outs by the operatives and a complete standstill is the result. National stoppages are extremely rare in the construction industry but in 1972, in order to support a wage claim and better conditions of employment, the operatives union, UCATT, selected to picket key sites throughout the country. These sites were large and of a multi-million pound nature, the result being that the employers acceded to the union's demands. There was obviously a substantial drop in the output of the industry during the strike period.

Numerous other industries service the construction industry and any problems of output they may have jeopardises the work on many sites.

To understand the magnitude of the industry it is necessary to study the details under appropriate headings which are shown as follows:

1. Building output

As can be seen from Fig. 1.4.1, the value of output by the construction industry in 1983 of nearly £24 346 m. was greater than the 1975 figure of almost £13 000 m. Although the output has more than doubled, inflation, which is caused by high wages and costs of materials, must be taken into account. In effect, shortly after 1975, there was a considerable reduction of investment in construction work by the country as a whole. So, from the 'go' period of 1971 to 1975 there has been a 'stop' period, certainly from 1975 to 1985.

The 'stop' period can be attributed to many factors, but in particular the general world recession caused at first by the world energy crisis. Britain's ailments in part were due also to the wages and salaries spiral of the mid-1970s, increases in holiday periods, better employment conditions and benefits awarded through legislation – although many of these benefits have been eroded, which in 1984 led to the 'summer of discontent' by some of the more powerful unions.

While the previously described conditions benefited the employees, the extra costs incurred were paid for by raising the prices of British goods which made exports less competitive on the ever decreasing world market. Conditions have changed, however, in the mid-1980s which has partially rectified the decline in the value of salaries and wages in real terms, but an increase in the numbers of long-term unemployed because of new innovations and techniques which require less labour.

The charts shown in Figs. 1.4.2 and 1.4.3 show clearly the divisions of work and output within the construction industry. A study shows the fluctuations, the reasons for which can be traced to one or more of the previously described problems including weather and industrial stoppages on site.

Industrialisation shares largely in the output of the industry with that of traditional building methods. Precast concrete and prefabricated buildings compare, cost-wise, very favourably with those built using the traditional techniques of cast *in situ* concrete, steel-framed structures and brick or stone. In 1978 the dwellings (flats, maisonettes, houses) which were completed in Great Britain for Local Authorities and Housing Associations, etc. was 136 600 – this was 48 per cent of all dwellings completed that year. The numbers have since reduced to 52 800 in 1982, as can be seen from Fig. 1.4.2.

The reason for the decline in new public dwellings has been caused by the Government's policy of believing in private ownership and the selling off to occupants of the public housing stock. Even private dwelling construction has declined because of the lack of demand. See Fig. 1.4.2.

Contractors' work is wide and varied and they compete for work not only with each other but sometimes bid for public work against the competitive prices submitted by various local authorities' direct labour departments. The type and level of work is shown in Fig. 1.4.3.

2. Material prices

These have risen at various rates depending on how scarce the materials became. It is interesting to note that some materials, say, copper, have risen at a reducing rate compared to bricks, the reason being that because of its scarcity and high costs during the 1960s, other materials emerged to take its place, particularly with regards to piping for water supply. Plastics and stainless steel tubes are now used by many in the industry.

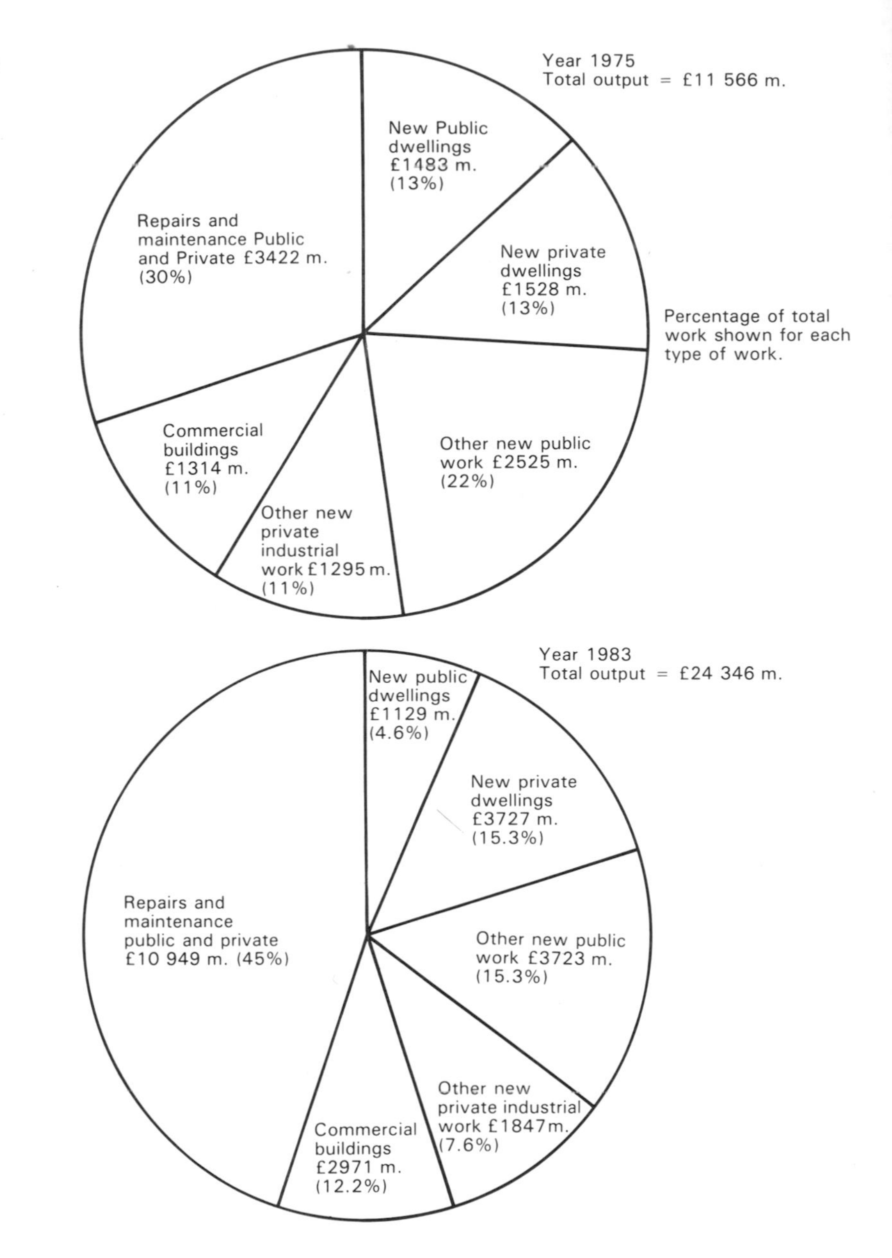

Fig. 1.4.1 Construction industry output (by type of work)
Note: Considering the level of inflation from 1975 to 1982, the value of new public dwellings constructed in 1983 has declined drastically. It is however interesting to note repairs and maintenance is approximately 45 per cent of total construction work

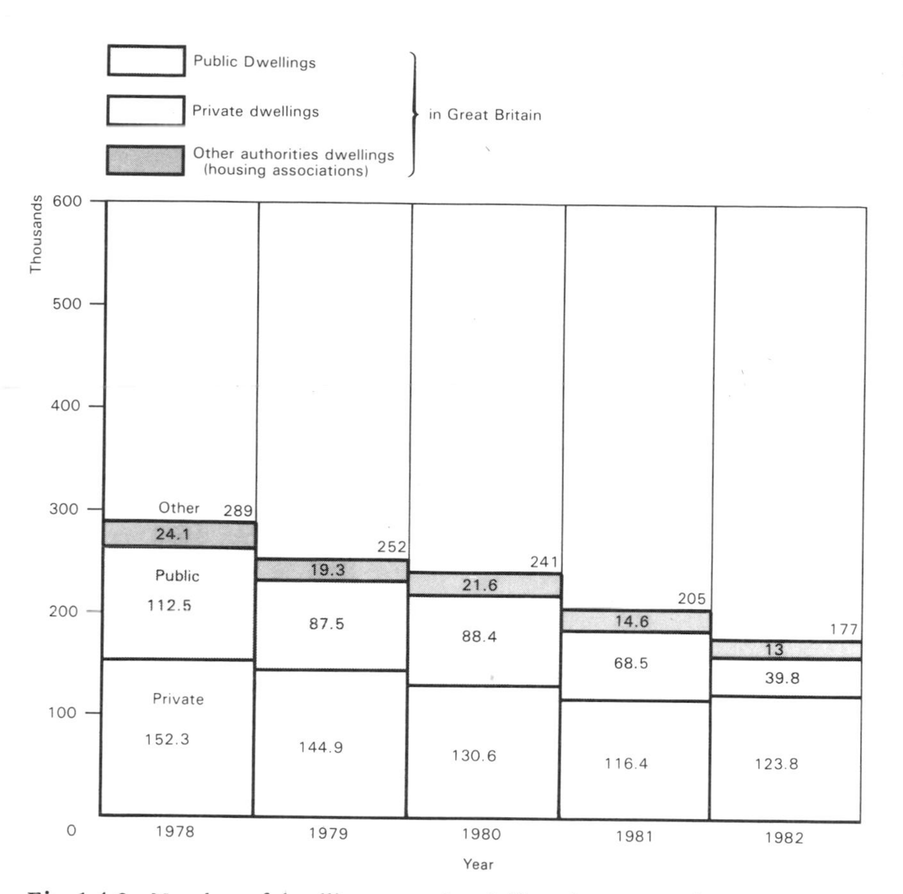

Fig. 1.4.2 Number of dwellings completed (flats, houses, etc.)
Note: There has been a substantial drop in the number of public dwellings built from 1978 to 1982 caused by the Government's policy of selling off housing stock to occupiers and believing in private ownership. Even then private dwellings completed have declined

One of the biggest factors which has led to substantial increases in manufactured materials was the astronomical rises in prices introduced by OPEC (Organisation of Petroleum Exporting Countries) brought about by their realisation that their stocks of fuel (namely oil) will not last for ever, so they restricted their output to make the commodity scarce and raised prices by up to 200 per cent. The industrial world soon felt the impact of this action because large quantities of fuel was, and still is, needed for production. Between 1980 and 1983 prices rose by $33\frac{1}{3}$ per cent. Figure 1.4.4 shows the varying levels of material price rises.

3. Wages

In the period from 1972 to 1975 wages and salaries escalated at a rate and to a level unprecedented before in this country. They rose approximately 163

Public and private sector

No.	Key	Type of work	Value in £ millions						
			1978	1979	1980	1981	1982	1983	1984
1		New dwellings (private and public)	3231	3321	2702	2685	3912		
2		Offices, shops, factories, garages	3190	3795	4098	4152	3980		
3		Gas, electric, water, sewage and coalboards	510	569	687	771	611		
4		Road, rails, air transport, harbours	652	655	832	1151	947		
5		Education, private and public	292	391	374	329	299		
6		Entertainment	248	308	342	288	282		
7		Health	261	287	320	478	467		
8		Miscellaneous	545	699	757	767	847		

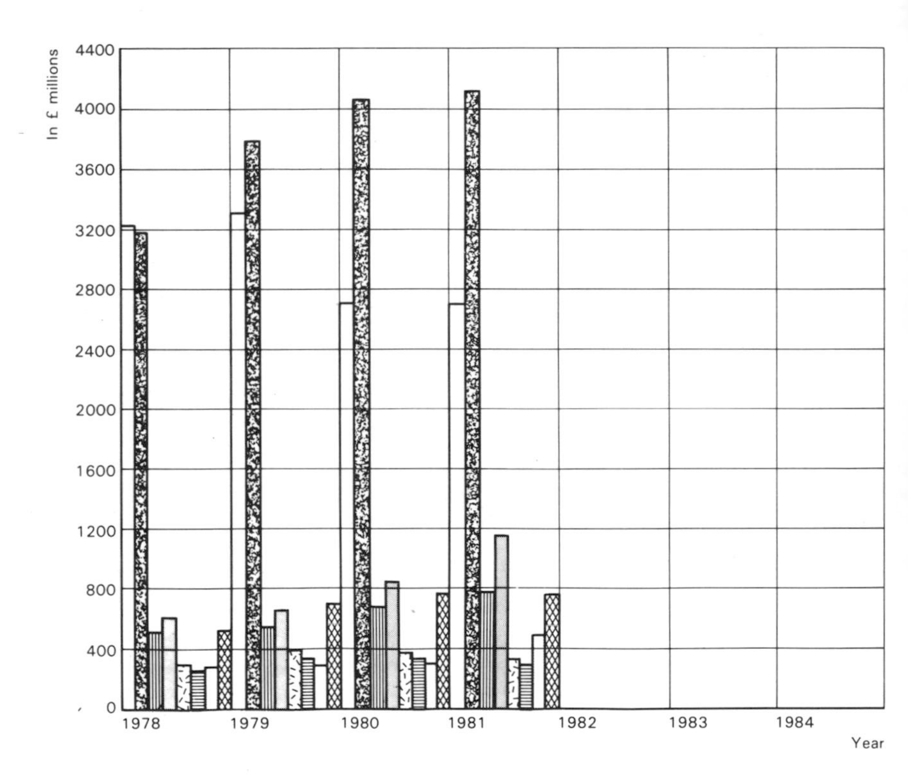

Fig. 1.4.3 Value of new orders obtained by contractors in the public and private sectors (at current prices). The figures do not include value of work undertaken by local authority direct labour departments or very small firms and self-employed workers

per cent. There had been rises due to the introduction of the Selective Employment Tax (SET) in 1968, by the Labour Government. This tax was levied on most firms in industry to bring in more revenue for the Government. Employers had to pay, at first, a tax of approximately £1.50 for each employee, so many employers in the construction industry found it cheaper to dismiss their operatives and re-employ them as labour-only subcontractors. This eventually became known as 'The Lump', a name which was adopted after the production of a television programme on the subject at that time.

In sacking and then re-employing the operatives on 'The Lump', the contractors were able to reduce their overheads, thereby incurring considerable savings on national insurance contributions, pension schemes, holiday stamps (public and annual), travelling expenses, lodging allowances, extra payments for dirty and uncomfortable conditions. There were even savings by not providing protective clothing and power tools on site by some contractors.

Labour-only subcontractors realised very early that in order to earn a living wage, after allowing for tax, insurance and other essential expenses, which they had to pay themselves, it was necessary to claim twice the amount already being earned. Because of the ever increasing shortage of skilled operatives, the employers began to compete with each other for the limited number of craftsmen which were available, but at a price. The craftsmen began to hold out to the highest bidder. There was such an acute shortage of certain key men that unscrupulous individuals began to pose as trained operatives, and after purchasing the minimum of tools were given employment by many contractors without any questions being asked. This situation was short lived but substantially added to the cost of construction work which is still at a high level even today.

4. Employees in the industry

In 1970 there was 1½ m. persons employed by contractors and public employers in the building and civil engineering industry, but this figure did not include architects, quantity surveyors, design specialists or construction workers in gas, water and electricity undertakings. The total figure therefore can be estimated to be around 1¾ m.

In 1983 only 1 090 000 were employed by contractors and public employers which shows a downward trend attributed to many factors, namely: fewer employees being required because of the reduction in construction work; new techniques, such as industrialised building which requires fewer operatives on site; more regular use of plant which leads to the reduction of manual methods; and because many people think the industry is too unattractive because of climatic and general working conditions. There are 20 m. people employed in all industries which means that there is approximately 5 per cent employed in construction.

To cope with the drop in numbers of trained operatives the Construction Industry Training Board (CITB) helps to recruit and train trainees of all ages.

High grants are sometimes made available from the CITB which encourages employers to take on apprentices (many firms do the training because they desperately need specialist tradesmen). The CITB look upon this method as an investment for the future, to help meet the demands for labour which they anticipate will be made on the industry in later years. The most common method of recruitment for apprenticeships at the present time is through the Youth Training Scheme (YTS) of the Manpower Services Commission (MSC).

Construction firms are charged a CITB levy, part of which can be claimed back through a grant when employees are sent on recognised courses or training schemes to help keep the industry's employees up to the required trained standard.

The Government's MSC contributes financially towards training within the construction and other industries, and most first and second-year trainees/ apprentices are paid through the YTS introduced by the Government in the early 1980s.

5. Construction firms

The numbers have decreased over the past ten years due to amalgamations or mergers, and by firms being forced out of business by lack of work. The number of recorded firms in 1970, being either one-man businesses or employers of a large labour force, was 73 420. This figure does not take into account the many self-employed craftsmen who signed on the dole and then worked on 'The Lump', receiving cash from main contractors without tax and other deductions, thereby swindling the Inland Revenue and reaping unjust rewards from unemployment benefits. This cost the country, it was estimated, about £100 m. per year. The Government has now partially plugged this gap by introducing a Registration Scheme for all labour-only subcontractors.

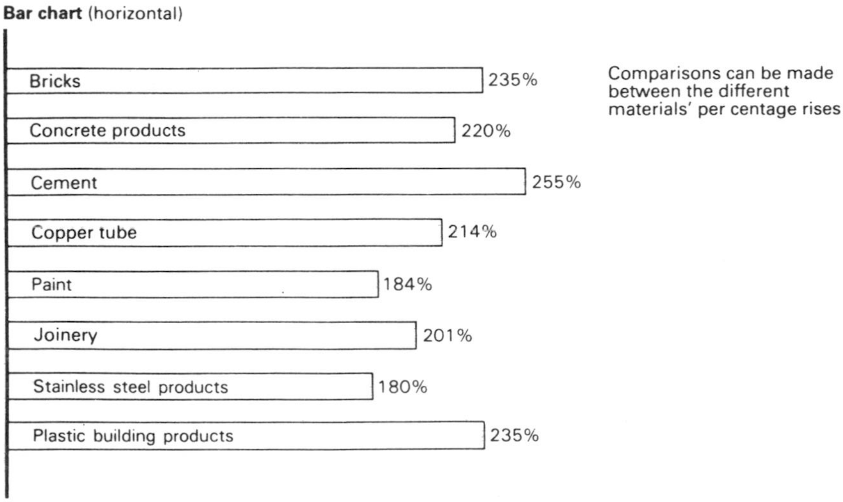

Bar chart (horizontal)

Bricks 235%
Concrete products 220%
Cement 255%
Copper tube 214%
Paint 184%
Joinery 201%
Stainless steel products 180%
Plastic building products 235%

Comparisons can be made between the different materials' per centage rises

Fig. 1.4.4 Material price rises of 1983 compared to 1980

6. Accidents in the industry

In 1983 the number of fatal accidents of operatives employed in building and civil engineering work was 131, which is one of the lowest recorded figures. It is anticipated that numbers will drop further in future years due to the 'Site Safe 83' drive which took place in the industry in 1983 and which is expected to be an ongoing campaign for a number of years. Other reported accidents also appear to have reduced in number and stand at approximately 22 000 for the same year.

If one considers those accidents which were reported – which necessitated some loss of production time – and those which were not reported, the figures must at least be double that shown in the Annual Report of H.M. Chief Inspector of Factories 1981.

Accidents cause distress to all concerned, and apart from slowing down production, result in the loss of earnings by the employee, and expenses are incurred by the employer, particularly with regards to extra insurance premiums which may have to be paid out later to insurance companies if they judge the firm to be a bad risk. Accidents also create bad publicity for the employer.

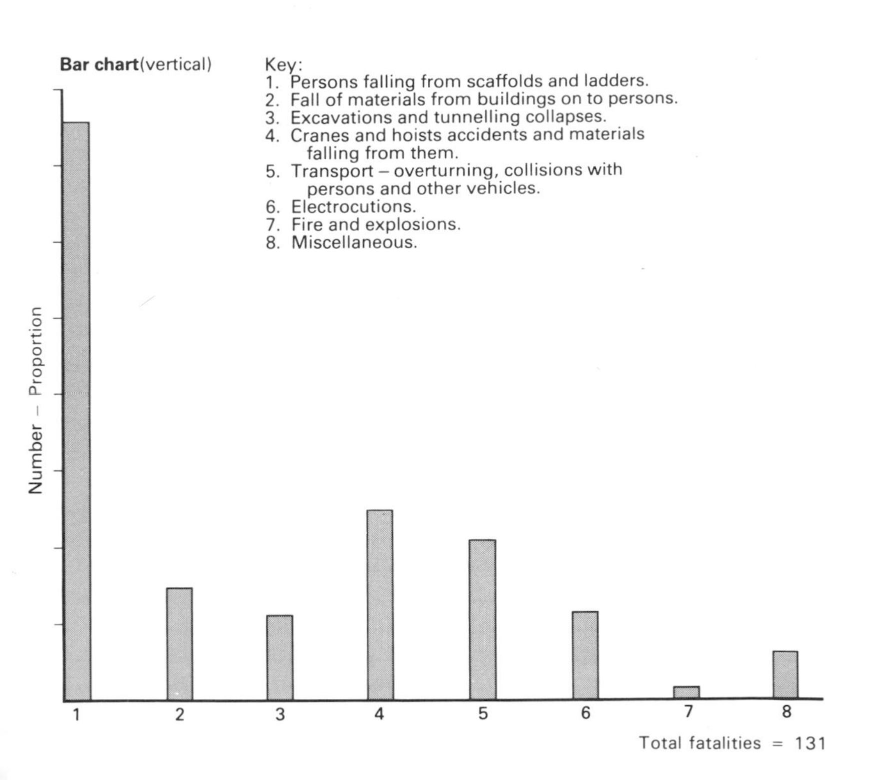

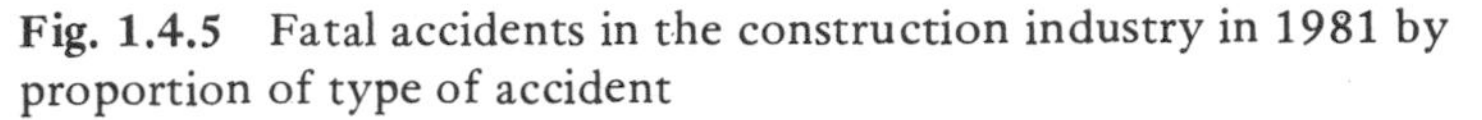

Fig. 1.4.5 Fatal accidents in the construction industry in 1981 by proportion of type of accident

The Royal Society for the Prevention of Accidents (RoSPA) helps to highlight risks, and contributes to all industries by running safety competitions, and prints other indoctrination leaflets and literature regarding safety.

The average number of fatalities over the past ten years in construction is about one per working day although of recent years there are about three per week. The fatal accidents for 1981 are highlighted, by type of accident, in Fig. 1.4.5.

1.5 Formation of a business and types

To form a business in the construction industry requires little or no capital. A person, usually a tradesman or skilled operative, seeks a client who needs a wall built, door hung or automatic washing machine plumbed-in, and on being successful to this end, is in business.

It is essential that there is continuity of work, so this means that further building activities must be lined up for the trader (as he would now be called) as soon as he finishes one job, in order to earn sufficient to support himself and his dependents. A full order or contract book obviously means that the business will eventually succeed, provided that the trader can meet the demands made upon him by starting and completing the work at the dates and times specified. Sometimes he may have to call upon others to assist him, either by employing them directly, or by subcontracting the work out in order to meet the deadlines imposed on him by the clients.

The minimum capital outlay for the types of work described would be in the provision of tools and materials. Later, a mixer, power tools or certainly a vehicle would be purchased to transport himself and his employees. Goods would also be purchased so that business can be conducted more efficiently. Care must be exercised against overspending on fixed assets (as they would be called) at the expense of materials and wages. One should keep a happy balance between these forms of expenditure and caution must be the keyword when first setting up in business.

Many one-man businesses successfully exist for an indefinite period, but there comes a time when others become more ambitious and the expansion of their trade becomes inevitable. More jobs and larger contracts are sought until eventually it matures into a large national firm.

The previously described business is known as a sole trader, and in many instances work as labour-only subcontractors. They can operate as subcontractors employing numerous operatives, or work in their own right as contractors tendering for work in a competitive situation. This trader is responsible for the day-to-day management of the business, possibly employing others to undertake the supervisory work on his behalf.

When a sole trader wishes or is forced to expand because of the possible lack of expertise or capital, he may invite others to invest their skills or money into the business and, as a reward, give them a share in the business management and the profits. This would usually be recognised as a partnership.

Company formation would be the next progressive step from a partnership, but a company can also be formed out of a sole trading business. Usually the company at first is registered as a private one, with the option of later expanding into a public concern. The types of businesses operating within the construction industry are therefore as follows:

1. Sole Trader.
2. Partnership. (*a*) Ordinary; (*b*) Limited.
3. Company. (*a*) Private (limited); (*b*) Public (limited).

Note: There are unlimited companies which do not normally operate in the construction field for reasons which become apparent later.

1. Sole trader

Formation

Anyone can commence trading in construction without having to register with the Registrar of Companies. One would normally have to register with the Inland Revenue, for pay-as-you-earn (PAYE); the Customs and Excise, for value added tax (VAT); and the Construction Industry Training Board, where the payroll exceeds the stipulated minimum limit.

Capital is usually required, whether it is in cash or kind, and is needed to carry out the work of the business or pay for any expenses. Assets, such as: office premises, office furniture, vans, lorries, materials in stock, and cash in the safe or bank is the capital of the business (less, of course, any liabilities, i.e. money owed).

The main problem with a sole trading business is that in the eyes of the law the proprietor (entrepreneur) has unlimited liability, and if debts accrue, then he is liable to his last penny. Some sole traders in their ignorance open two bank accounts, one for private use and the other being for the business, believing that if the business fails, the money in the private account cannot be called upon for any business debts. This is a wrong assumption in the eyes of the law, as all bank accounts belonging to the sole trader are classed as one for the paying of creditors. Even the personal belongings of the businessman could be used to satisfy those to whom he owes money.

An advantage of being a sole trader is that complete control of the business remains in his hands; also, any profits earned are entirely for his benefit. The disadvantages, on the other hand, are that losses are borne entirely by the one person, and skills and knowledge are limited; two heads being better than one, is the old saying.

As a sole trader it is important to know the following management terms:

(*a*) **Solvency,** being enough funds to pay off debts to creditors, therefore allowing the business to continue trading.
(*b*) **Insolvency,** being insufficient funds to pay off creditors, which eventually leads to the discontinuance of trading.
(*c*) **Bankruptcy,** where insolvency exists and creditors sue for payment in a Court of Law. The Court would usually declare a trader bankrupt, which

means that he will be unable to start up in business again until he can satisfy the creditors. Neither will he be able to hold a directorship of any other business, or hold public office.

(*d*) **Liquidation,** the winding up of a company due to insolvency. Companies do not become bankrupt, only individuals can become so.

2. Partnership

This is when a minimum of two persons combine with a view to making profits. The maximum laid down in law is twenty, but this may be increased for certain professional concerns. In the building and civil engineering industry, however, the former maximum figure must be observed if it is thought necessary to in fact create a partnership at all.

It is extremely unlikely that anyone would contemplate forming this type of business when the facilities offered, through forming a private limited company, are so advantageous for building and civil engineering contractors.

There does not have to be a written agreement between persons working togethei for them to be classed as a partnership. Usually, if the net profit earned through trading is shared out amongst persons trading jointly together, then in law it would be classed as a partnership. This, of course, could lead to disastrous consequences later if a substantial loss was made thereby resulting in debts being accrued. The creditors could sue for payment and if one partner is insolvent the other could be called upon to pay both shares of the debt. Naturally, this is one of the main disadvantages of a partnership, particularly ordinary partnerships.

The legislation which controls partnership businesses are:

(*a*) The Partnership Act of 1890, for ordinary partnerships.
(*b*) The Partnership Act of 1907, for limited partnerships.

(a) Ordinary partnerships

The formation of these concerns is similar to sole trading businesses and does not normally need registering. Similarly, each partner has unlimited liability which can be an unnecessary risk particularly in the construction field. It is therefore necessary to choose partners wisely and to be prudent enough to jointly approach a solicitor with a view to having a 'Deed of Partnership' drawn up. This document, when prepared, becomes a contract between the partners, and normally if prepared carefully by a solicitor specialising in this field, using as a guide Precedents which have been set in the past, should state clearly the rights and duties of each partner.

It should be stated here that it is most unusual for a firm of builders or civil engineering contractors to operate under a partnership, unless the business is a new one and the partners understand little about business structures.

In preparation of a Deed of Partnership by a solicitor, Precedents are used to cover all points which are relevant regarding the peculiarities of a partnership which is about to take place. The Precedents are judgements given in previous judiciary cases which act as a guide to future judgements of a similar nature. The peculiarities of one partnership compared to another can be illustrated by comparing the formation of a partnership in building between two persons, one of whom wishes that his share of the partnership should fall to his heir in the event of his death, and the other firm of, say, solicitors, where one invests more capital than the other and requires a larger share; also, he wishes to receive emoluments during the first five years of his retirement.

It can be seen that with these two firms the Deed of Partnership needs to be prepared with care, and that, essentially, 'Deeds' must be tailor-made to suit the requirements of the partners. Some of the clauses which would be incorporated can be outlined as follows:

1. Name and address of the firm. Usually the name incorporates the surnames of the partners, unless an existing partnership is purchased.
2. Type of business and extent of the work to be undertaken.
3. Partners' names and addresses, so that each partner knows exactly who is involved.
4. The duration of the partnership, which is usually for a fixed period but can be extended if required, or on the death of a partner.
5. Status of each partner within the business, whether one is a general or limited partner.
6. The method by which a partnership can be dissolved. Usually it is by the mutual agreement of all concerned.
7. Amount of capital invested in the firm and the share of the profits or losses.
8. Amount of interest on capital to be paid if necessary.
9. Maximum withdrawals of money from the firm for personal use, on a yearly basis, and the amount of interest to be charged for this privilege.
10. Partners' salaries or other methods of payment.
11. Interest to be paid to partners who lend money to the firm.
12. Type of books and accounts to be kept for management control.
13. Name of the auditors to be used and at what intervals.
14. Repayment of capital, and adjustments thereby for Goodwill, on partners' retirement or death.
15. Arbitration, and how this should operate in the event of the partners' serious disagreement (it is quicker than going to court and also saves on bad publicity due to the hearing being held privately).
16. Position of each partner, such as: general manager, accountant, sales manager.
17. Course of action to be taken when the question of unprofessional conduct arises.
18. Admittances of new partners; usually by unanimous agreement.
19. Non-competitiveness of partners. Partners should not belong to another partnership in similar line of business.
20. Limitation on partners. Work should not be contracted into which falls outside the scope of the business, and in other cases a maximum figure should be laid down.

Intending partners can have written into a Deed any points which are mutually acceptable to each other and which will safeguard against any misunderstandings arising at a later date.

If a Deed of Partnership is not prepared, or certain points are not covered within one which is prepared, then the Partnership Act of 1890 would apply, and a ruling given in a court of law would take into account that which is stated in this piece of legislation, which is:

- Profit and losses are to be shared equally.
- There should be no interest paid on capital.
- All partners must agree to new partners joining the firm.
- Internal differences in the running of the business are to be settled by a majority vote of the partners, except where there is to be a major change, then all partners must agree.
- Interest on loans by partners to the business can be 5 per cent.
- Each partner is entitled to invest the same amount of capital as other partners.

(b) Limited partnerships

In an ordinary partnership each partner has unlimited liability. Some persons, however, may be willing to invest money in the firm but do not wish to participate in its management. They may also want to limit their liability should the business fail. The answer would be to form a limited partnership giving these partners the status of limited or sleeping partners.

The way to change from ordinary to limited partnership would be as stated in the Partnership Act of 1907. Registration with the Registrar of Businesses (or Joint Stock Companies) should be done immediately, and while partners can have limited liability, at least one partner (known as a general partner) must have unlimited liability for the firm.

Types of partners

General partners: Can be one or more partners who have unlimited liability and, as such, have liabilities similar to a sole trader. They also take an active part in the management of the firm and are agents for the other partners.

Limited partners: Are also known as a sleeping or dormant partner having limited liability, that is to say, they can only be liable to the value of the capital invested into the business. They must not take an active part in the business affairs; if they did and the business became insolvent, they would be classified as general partners and would have unlimited liability.

3. Companies

The majority of companies are Incorporated by registering with the Registrar of Joint Stock Companies in London, but others, in very exceptional circumstances, can be incorporated by Royal Charter or by an individual Act of Parliament. Because of its incorporation (combined into one) a company becomes a corporate body, and as such in law it can sue or be sued similar to a person, or hold property in its own name. Unlike a sole trader or partnership, when the business is not a legal body, a company would still continue in existence in the event of a shareholder leaving the firm, or on the death or bankruptcy of a member.

The Companies Act 1948 (the principal Act), with the addition of the Companies Act 1967, 1976, 1980 and 1981, lays down the procedures to be followed regarding the formation and running of a business. There can be unlimited or limited liability companies. The former is rare, and in any case would not be appropriate in the construction and other industries because of risk involved to the shareholders. The limited liability companies which can be incorporated under the Companies Act 1948 are:

(*a*) Companies limited by guarantee, as in non-profit making concerns. For example: the various professional institutes, such as the Chartered Institute of Building, are typical companies limited by guarantee. Members are limited regarding their liability, in the event of the organisation becoming insolvent, to the amount they agree to subscribe annually for membership.

(*b*) Companies limited by shares, as in private or public trading companies. Shareholders are only liable for the company's debts to the value of the shares they hold.

It should be noted that the word 'limited' on a business name acts as a warning to others who trade with the company or organisation. The private assets of a shareholder is also safeguarded.

Under the details laid down in the Companies Act 1948, businesses limited by shares, which are operated to make profits as distinct from non-profit making concerns, are divided into private and public limited liability companies.

(a) Private limited companies

To register this type of business there must be at least two shareholders. During the existence of the company the number of shareholders must not exceed fifty, discounting past and present employee shareholders.

If a business is to remain as a family concern giving control to those who started the business (similar in nature to sole traders and partnerships) and at the same time restricting the liability of its members, then a private limited company is the answer with the word 'Limited' or 'Ltd' after the name.

There are other distinguishing features of a private company compared to those of a public company, which are:

(*a*) Shares cannot be sold on the open market by advertising, but are usually offered to relatives, friends or employees.
(*b*) The firm's articles would normally insist that when shares become available for sale they must first be offered to existing shareholders.
(*c*) On registration, the company can commence trading immediately.
(*d*) Statutory meetings need not be held, neither should it be compulsory to submit a preliminary statutory report to the Registrar.

(b) Public limited companies

A minimum of two (instead of the usual seven) shareholders may form a public company (Companies Act 1980) and there is no maximum limit. In numerous instances, these businesses are created by converting from a private limited company by registering the fact with the Registrar of Joint Stock Companies, especially when more capital is required to expand its business operations.

The important features of a public limited company are:

(*a*) Shares can be offered on the open market either by advertising or through a stockbroker.
(*b*) Existing shares can be bought or sold at the Stock Exchange, thereby, allowing freedom of transfer.
(*c*) When a new company is formed a trading certificate is required from the Registrar of Joint Stock Companies before trading can commence.
(*d*) The letters PLC (public limited company) should suffix the company name.

Formation of a company

A written application, with the required registration fee, must be made to the Registrar of Joint Stock Companies if incorporation is required. It should be signed by the applicants/founders, the minimum permitted number being two for private and for public companies. Although a private company can commence trading immediately after submitting documents to the satisfaction of the Registrar, the public company must wait until sufficient capital has been raised by the issuing of shares, before being eligible for a Trading Certificate.

Registration can be made in London, Edinburgh or Belfast, depending on where the company wishes to have its registered office.

The founders of private companies must prepare the following main documents:

1. Memorandum of Association.
2. Articles of Association.

Also, in the case of a public company, further details must be submitted which are:

3. A list of persons willing to be directors.
4. Written consents of each director, which is duly signed by them.
5. A signed statutory declaration by a director and the company secretary that the Companies Act 1948, 1980 and 1981 has been complied with in every respect.

On satisfactory returns received by the Registrar, a Certificate of Incorporation would be issued making the company a separate legal entity. This allows it to trade in its own name separate from the shareholders. It can enter into contracts, can own assets and be in its own right subject to the law of the land.

Memorandum of Association: This is a contract between the company and those from outside the business with whom it trades. It is assumed that everyone knows the contents of the Memorandum of Association when undertaking trade with the company.

The Companies Act 1948, 1980 and 1981 lays down that the Memorandum of Association must include the following five points:

1. The name of the company, with the last word 'Limited' included for private companies and PLC for public companies.
2. The registered office, whether in England and Wales, Scotland or Northern Ireland.
3. The objects of the company, and type of business.
   If the directors then decide later to contract for work of a different nature to that shown in the Memorandum, the shareholders could seek an Injunction against the directors in a court of law forbidding them to do so.
4. State that the liability of its shareholders is limited.
   The Act states, however, that if the number of shareholders falls below two in both a public company and a private one, and remains so for six months or more, the liability then becomes unlimited.
5. The amount of share capital with which the company is expected to commence trading.

Articles of Association: This is a contract between the company and its members and contains the regulations and rules for conducting business within the company.

If Articles of Association are not prepared, it is assumed that the Table A Model Set of Articles contained within the First Schedule of the Companies Act 1948 apply.

In the event of serious disputes within the company, the Memorandum takes precedence over the Articles. The Articles however may include the following clauses:

1. The authority, duties and rights of the director and other officers and members.
2. Voting rights of shareholders.
3. Procedure for calling and conducting meetings.
4. Method of appointing officers.
5. Payment of directors and methods by which their salaries can be reappraised.

Before a public company can start trading, on being incorporated, sufficient shares must first be sold to show that the company has a reasonable chance of success. The total value of the shares and the minimum subscription is laid down in the Prospectus, and when shares are offered for sale the minimum subscription must be reached before the company is given a Trading Certificate. Undersubscription would make the applications void and the money received by the company would have to be returned.

When offering new shares for sale, as compared to old ones which are normally sold through the Stock Exchange, one of two methods may be used:

1. A Prospectus.
2. An Offer for Sale.

A prospectus

Subscription for shares can be offered to the public (public companies only) through a prospectus, a copy of which is sent to the Registrar of Joint Stock Companies. This document gives details of the business and its proposed field of activity. It is a means of inviting the public to subscribe for shares or debentures by advertising through a newspaper, notice or circular. It would, amongst other things, include the following:

1. Names and addresses of directors.
2. The name and type of business of the company.
3. The capital required to start business (better known as the Minimum subscription).

An offer for sale

Stockbrokers, merchant banks or issuing houses may be used as 'agents' by the company's founders to circulate the prospectus for the issuing of shares to the public, but sometimes an issuing house buys all of the shares and then proceeds to sell them to the public by circulating an 'offer for sale' which in effect is similar to a prospectus. The profit, if any, from the sale of shares this way will belong to the issuing house. This usually means that the shares will be sold at a premium, as compared to a discount when selling for less than the shares' face value.

When issuing shares the company may, in addition to issuing them to the public generally, issue them privately to persons known to the founders or other shareholders. Also, existing shareholders could be 'of right' offered the shares if the company is already in existence, and there is to be a new issue when further capital is required.

Types of shares

Various shares exist within companies and each type have distinctive features, particularly when the profits of the company are to be shared out in the form of dividends.

There are two types of shares:

(*a*) Preference shares.
(*b*) Ordinary shares.

(a) Preference shares

Ordinary preference shares: A fixed dividend is payable on this type of share, and is taken as a percentage of the share's nominal value. Preference shares have first claim on profits made by the company, but in the event of excess profits being made would not normally share in the surpluses being distributed. If there are no profits in a particular year this type of share would lose in the same way as ordinary shares. Usually, preference shares have no voting rights.

Cumulative preference shares: They are similar to ordinary preference shares but if the profits in any one year are insufficient to pay the fixed percentage dividend, then this value is carried to the next year and would be payable along with the subsequent years' dividend.

Redeemable preference shares: These can be issued in accordance with the Companies Act 1948 and have the same rights as ordinary preference shares. The company, however, has the right to redeem them (buy back) in the future if it so requires.

Participating preference shares: Gives the right to participate in any surplus profits in addition to the fixed percentage normally paid.

(b) Ordinary shares

The majority of shares issued by any company are of this type. The characteristics of ordinary shares are:

1. Full voting rights are given to holders of these shares. The more shares being held by the individual, the larger his/her vote will be at shareholders' meetings. If over 50 per cent shares are held by a member of the company, that member would, in effect, control the business.
2. There is no fixed rate of dividend as in preference shares. The dividend is announced at the Annual General Meeting (AGM).
3. Ordinary share dividends are paid after preference shareholders receive their dividend.

The selling of shares is only one way in which a company can raise capital. There are many more ways, as described previously in Section 1.2, along with the issuing of debentures.

Debentures

When funds are required by the company Debenture certificates can be issued to individuals or organisations in exchange for the loan of money either for a long or short term period. Normally, the debentures are redeemable and give a satisfactory rate of interest to the holders in exchange for the use of their money. They are a sound investment as they usually allow first claims on the company's assets when it gets into financial difficulty and is forced into liquidation.

Shareholders are part owners of the company, but debenture holders are creditors and therefore have no voting rights, but, whether profits are made or not, debenture holders still receive their fixed rate of interest.

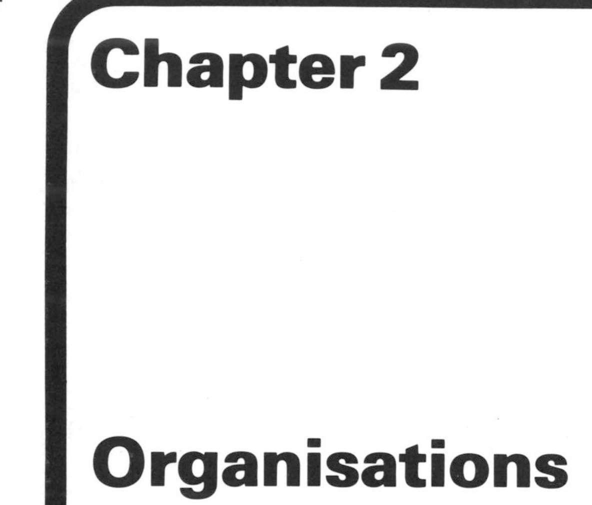

2.1 Professional and other institutions, including societies

If standards are to be maintained in the Industry it is essential that those associated with construction must be provided with knowledge and mental skills. Suitable education to this end is therefore necessary and is divided into two parts:

1. Technical Education.
2. Practical Education.

The former requires some form of study at a Higher Education Establishment comprising either a University, Polytechnic, College of Technology, College of Further Education or Technical College in a full or part-time capacity. For Practical education the trainee is required to work under the supervision of qualified and trained personnel, who have gained experience in the type of work the trainee wishes to pursue.

On completion of the desired Technical and Practical education relevant to the vocation being pursued, a trainee may be eligible to gain full membership of an institute which is most appropriate to his/her standing in the Industry.

The qualified people of the Industry are usually placed in one of three categories, which are:

1. Professionally Qualified.
2. Technicians.
3. Tradesmen or Craftsmen.

Those who hold Corporate membership of a professional institute normally develop executive abilities, and eventually fill the important managerial positions in the Industry. The Technicians, on the other hand, fill the middle management posts and direct the operations being undertaken by the craftsmen.

Professional Institutes

There are numerous Institutes having members employed as accountants, administrators and legal advisers within construction firms, but only those which are predominantly aligned to the Industry are dealt with here. The following arc important:

The Royal Town Planning Institute (RTPI) – 1914

This Institute's members are either employed directly by national or local Governments, or they practice as professional consultants.

The circumstances leading to the development of this Instititute, was due to the Industrial Revolution and the subsequent flow of agricultural workers to towns and cities. In some areas of Great Britain the people were exploited by the Industrialists and lived in poverty, consequently many Humanists emerged and strove to improve the living conditions of these people.

Consideration was given on how improvements could be made to public health, housing and the juxtaposition of dwellings. A healthy environment was considered necessary and the siting of buildings being of paramount importance. One such Humanist built a whole village in 1879 for his own employees in pleasant surroundings. His name was George Cadbury and his village was called Bourneville, near Birmingham.

Names such as Shaftesbury, Owen, Lever and Saltair are associated with championing the cause of the exploited, and were instrumental in the passing of Acts of Parliament which eventually led indirectly to the Town and Country Planning Acts of the twentieth century.

The introduction of the Housing, Town Planning, etc. Act 1909, which empowered local authorities to control future development prevented such things as Ribbon and other unsuitable development. The formation of the Royal Town Planning Institute followed, and now has the following levels of membership: Member (M), Associate Member (AM), Legal Member (LM) and Legal Associate Member (LAM).

The Royal Institute of British Architects (RIBA)

This Institute received its Royal Charter in 1840 and since then has increased the status of its members in society.

Very few Architects undertook formal training until the eighteenth century; in fact, many had worked in different professions until they had reached middle age. Two typical Architects were: Inigo Jones and Christopher Wren. Inigo Jones was originally a scenery designer, whereas Wren was an Astronomer and Mathematician at Oxford University in the seventeenth century.

Many of the earlier designers of buildings worked in the Building Industry for many years before graduating, by experience, from journeymen to master builders, and then into architectural design work. The early Architects by the very nature of their training had a very sound knowledge of their materials and construction, and their knowledge was enhanced by their many years' experience.

It has shown through the years that an Architect must have, amongst others, the following attributes:

1. Foresight – so that his/her building remains aesthetically satisfactory.
2. Artistic ability.
3. An understanding of materials – so that he must be a scientist and technologist.
4. An ability to communicate – be able to interpret the client's needs and convey them to the builder.
5. Coordinating ability – of all professional and other personnel in the design and construction team.

Courses in Architecture are available at some Universities and Polytechnics and a limited number of students are admitted on these courses each year.

Entry qualifications are high, and only those who have obtained Advanced Level School Certificates are considered. When students graduate from College with a Degree in Architecture on the conclusion of three to four years' study, they must then satisfy the RIBA membership requirements regarding practical experience and competence before being admitted to Corporate Membership. This entails serving in an Architect's office, after qualifying academically, usually to the age of twenty-five. They will then be eligible for Associate Membership (ARIBA).

The Royal Institution of Chartered Surveyors (RICS)

This Institution was founded in 1868 and obtained its Royal Charter in 1881. The Institution is divided into six sections of which the General and the Quantity Surveying Sections are primarily concerned with existing and new building work. The remaining four are: land surveying, mining, agricultural and hydrographical sections.

Students may qualify at the Further Education Establishments either by attending a degree or full/part-time RICS course. Two appropriate Advanced Level School Certificates or other qualifications acceptable to the Institution is necessary for entry into any course if students are to be successful in their application for membership of the RICS at a later date.

Members of the RICS usually work in a professional capacity. They may have their own Practice; are employed by a Partnership; or are Professional

Grade employees of a local or central Government. There are many, however, who are directly employed by building or civil engineering contractors.

The Institution of Civil Engineers (ICE)

This is the oldest of the Professional Institutions and dates back to the year 1792. Important names associated with this Institute are: John Smeaton, the famous Lighthouse Designer, and Thomas Telford, who has many great civil and structural achievements to his credit.

The Institution was granted a Royal Charter in 1828. Corporate members are not only allowed the designated letters of MICE (Member) after their name, but that of Chartered Engineer (CEng). This Engineering Institution, like many others, is a member of the Council of Engineering Institutions (CEI).

The CEI is a federation of over fifteen engineering institutions, and on being granted a Royal Charter in 1965 gave it the authority to grant the title of Chartered Engineer to Corporate Members of the member institutions.

To qualify for full membership (Corporate) of the ICE, students/pupils/ trainees must normally have attended full-time or block-release courses at a University or Polytechnic studying for a degree.

The work undertaken by a fully qualified Civil Engineer (Civilian Engineer as compared to the Military Engineers of 'bygone' years) is usually in designing and supervising the construction of roads, airports, railways, bridges and large Government projects.

It is expected that the new Engineering Council will take over from the CEI.

The Institution of Structural Engineering (IStrucE)

Originally, when it was formed in 1908, the subject of its members' work was to deal exclusively with the use of Reinforced Concrete and Concrete Technology. The work rapidly expanded to include other materials such as wood and steel.

The Institution has a Royal Charter, and to obtain Corporate Membership, post Engineering degree experience is required in professional practice, and persons must reach a minimum age of twenty-five years.

Many members of this Institution have their own, or are employed in, consultancy practices. They are called upon by Architects and Designers to calculate sizes of structural elements and components of a building, such as: foundations, walls, columns, floors and roofs. This often requires them to undertake immensely complex calculations. They also work in their own right designing bridges, retaining walls and other structures which are of a complicated nature.

Other Engineering Institutes

During the design stage the Architect may invite other Specialist Consultants to give advice on the internal environment of the proposed building. Separate drawings could be prepared showing the different services which need to be installed to improve living conditions within the new-built environment. Plumbing, drainage, lighting, heating, ventilation, sound insulation, thermal insulation, telephone systems and lights are usually the more important types of installations Consultants help to design.

The Consultants or Assistants to the Architect who deal with the services design may be members of the following bodies:

The Institute of Electrical Engineers (IEE).
The Institute of Mechanical Engineers (IMechE).
The Institute of Public Health Engineers (IPHE).
The Chartered Institute of Building Services (CIBS).

A substantial number of employees of Local Authorities and Public Health Undertakings working in the field of surveying and engineering are members of the Institute of Municipal Engineers.

The Chartered Institute of Building (CIOB)

It was formed as the Builders' Society in 1834 but was later Incorporated in 1884 and was recently granted a Royal Charter in 1980.

There are various levels of membership of this Institute but the two Corporate grades are Members (MIOB) and Fellows (FIOB). Full voting rights are restricted to the Corporate members who have mainly qualified by satisfying the Institute regarding the following:

(*a*) A pass in the Institute's Final examination.
(*b*) Two years' post Final examination experience in a responsible position.
(*c*) Success in the Institute's professional interview.

The various educational routes to Corporate Membership of the Institute is outlined in Fig. 2.1.1.

The members of this Institute are associated with a wide range of business and professional interests. These interests lie in the ownership of contracting businesses, and many are directors, managers, supervisors, surveyors and lecturers.

Various selected members act as representatives of the CIOB on committees of the Construction Industry Training Board (CITB), the British Standards Institute (BSI), and committees made up of other professional institutes which have mutual interests in building and civil engineering.

The CIOB plays a prominent role in most aspects of the Industry and makes representations to the Government regarding proposed legislation.

There are nearly 27 000 members of this Institute, of all grades, from Student to Honorary Fellows. It is therefore a major contributor, in every respect, to the Industry.

The levels of membership of the CIOB are: Student, Associate, Licentiate, Graduate, Member (MCIOB), Fellow (FCIOB), Honorary Fellow (HonFCIOB) and there are Retired members.

The British Institute of Management (BIM)

Many members of the aforementioned Institutes enjoy membership of the BIM. The advantages being, that through this Organisation's publications and meetings the academics/scientists/technologists can study current trends in administration and management. It is an institute where members of the whole spectrum of industry and commerce can meet to discuss and compare problems and ideas. Those who are principals of the organisations to which they belong usually progress to the BIM Member grade (MBIM). Those who are middle managers or supervisors may reach Associate Member level (AMBIM).

The Construction Surveyors Institute (CSI) – 1952

Members are normally employed in building, civil engineering, mechanical and electrical engineering, and other allied industries. Also, some are employed in Municipal offices in departments of architecture and surveying.

For Corporate Membership grade of Member and Fellow requires a degree level qualification. The Ordinary and Higher Technician Certificate is the more usual route to the full membership level. (At present there are negotiations taking place with the Chartered Institute of Building with a view to a possible merger.)

Incorporated Association of Architects and Surveyors (IAAS)

This was founded in 1925 and embraces many of those who are employed as quantity surveyors, municipal building surveyors, general surveyors, land surveyors, fire surveyors, valuation surveyors, town planning surveyors and security surveyors.

Technician institutes/institutions

After the formation of many of the professional institutes it was evident that for those not professionally qualified, but in the main were employed by professional people, organisations were necessary to maintain their standing in the Industry. Some of the professional bodies played their part in helping to sponsor the Technicians' Institutes. They were formed primarily to ensure that satisfactory standards were maintained in technical ability and general competence by their members.

Corporate Membership of an institute indicates the intellectual ability of a member, and shows that satisfactory experience has been gained within the appropriate sphere.

Monthly or periodical journals or papers of the institutes help to inform members of new developments in their field of interest. Members must of course be willing to contribute articles to the journal/paper for its production to be a success.

There are a number of professional institutes which allow Technician Membership in addition to specially formed Technician Institutes, the Chartered Institute of Building being undoubtedly one of the more important ones.

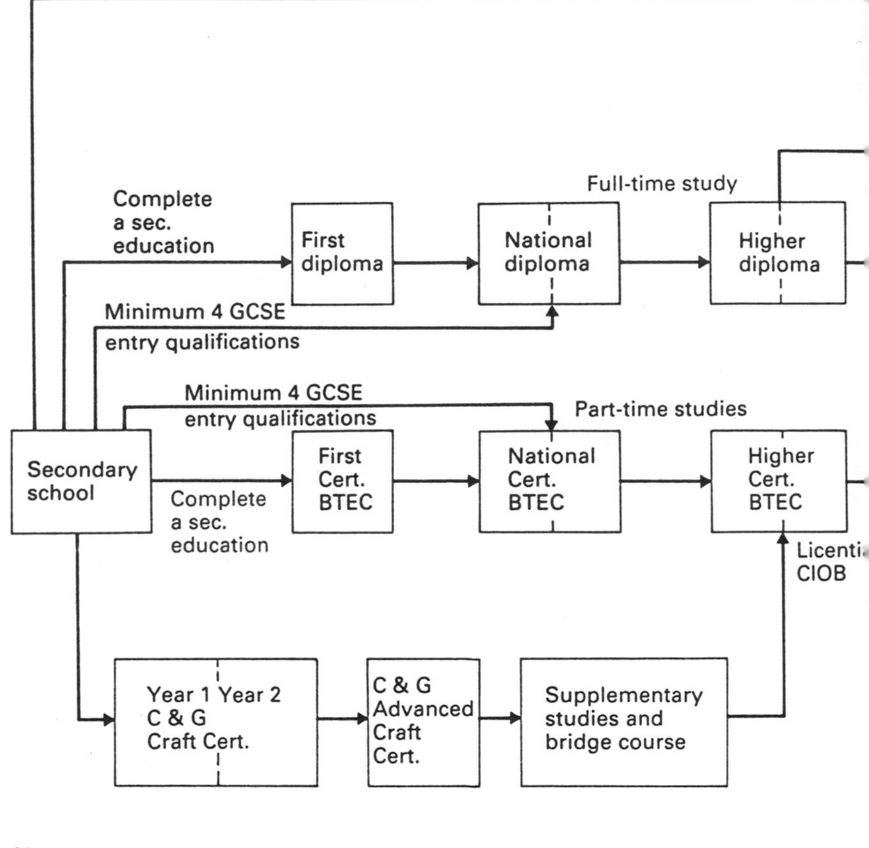

Note:

C & G City & Guilds of the London Institute

These diplomas/certificates are administered by BTEC (Business and Technician Education Council)

BA/MA Bachelor/Master of Arts
BSc/MSc Bachelor/Master of Science
PhD Doctor of Philosophy
[ ] Designated CIOB class of membership

Fig. 2.1.1 Educational routes to Corporate Membership of the Chartered I

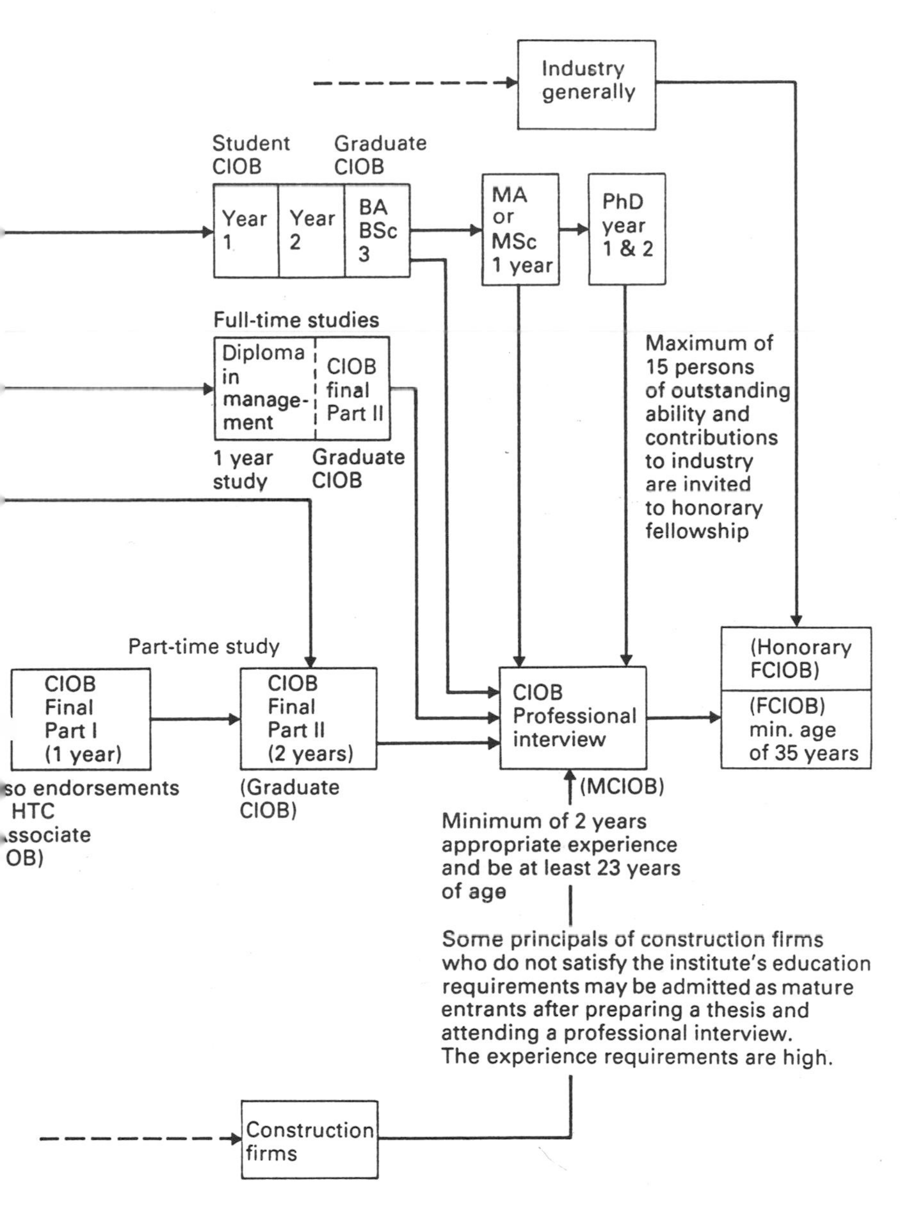

Industry generally
Student CIOB
Graduate CIOB
Year 1
Year 2
BA BSc 3
MA or MSc 1 year
PhD year 1 & 2
Full-time studies
Diploma in management
CIOB final Part II
1 year study
Graduate CIOB
Maximum of 15 persons of outstanding ability and contributions to industry are invited to honorary fellowship
Part-time study
CIOB Final Part I (1 year)
CIOB Final Part II (2 years)
CIOB Professional interview
(Honorary FCIOB)
(FCIOB) min. age of 35 years
so endorsements
HTC
ssociate
OB)
(Graduate CIOB)
(MCIOB)
Minimum of 2 years appropriate experience and be at least 23 years of age
Some principals of construction firms who do not satisfy the institute's education requirements may be admitted as mature entrants after preparing a thesis and attending a professional interview. The experience requirements are high.
Construction firms

British Institute of Architectural Technology (BIAT) formerly the Society of Architectural and Associated Technicians (SAAT)

The RIBA promoted this Technicians' Institute in 1965 to help govern the standards of training and education for architectural and other technicians. A HTC (Higher Technician Certificate) qualification and appropriate experience for the Membership Grade is essential.

The Society of Surveying Technicians (SST) – 1970

The usual route to Member grade (MSST) is by studying to the HTC/HTD level in any of the following disciplines: building, estate management, surveying and cartography, mining or agriculture. Intending members must also undergo formal training in the same subject being studied.

The Society of Civil Engineering Technicians (SCET)

It was recognised that the Technician employed by civil engineers or firms of civil engineering contractors undertook work of a very important nature, and because of this their employers found it necessary for the individual to undergo formal training, and be allowed to attend higher education classes. On qualifying with an OTC/OTD–HTC/HTD in civil engineering the trainee may apply for Registration as a Technician or Technician Engineer under the scheme administered by the Council of Engineering Institutes (CEI). On registration each trainee (now qualified) becomes eligible to join the SCET.

The work undertaken by civil engineering technicians is similar to that of the professionally qualified civil engineer but in a subordinate capacity.

The Institute of Clerks of Works of Great Britain Incorporated (ICW) – 1882

There are three levels of membership, the Licentiate, Associate and Fellow. These levels may be gained by success in the Institute's examinations or by equivalent qualifications.

Members work for clients of construction whether they are in the private sector or public sector. They assist the Architect in the supervision of the on-site work. Architects often retain the services of a Clerk of Works and the client accepts this as a very essential part of the services offered by the Architect.

Inspections are normally carried out continually during the course of the construction work on the standards of workmanship of the contractor. There is also other administrative work undertaken on site by the Clerk of Works.

The majority of Clerks of Works have progressed through a building craft such as: carpentry and joinery, bricklaying, masonry and plastering to reach this important post of responsibility.

Craft institutes

These Institutes attempt to promote acceptable standards for the craft which they represent. Many employers acknowledge these standards and insist on filling vacancies within their organisations, especially where trade foremanship positions are involved, with qualified craftsmen who are also members of an approved Institute. This is understandable if the standards of workmanship is to be maintained.

The Institute of Carpenters (IOC)

This Institute was formed to advance the status of qualified carpenters and joiners, and to set high standards by compelling membership examinations for those who are not qualified to the First Class level of the City and Guilds of London Institute Advanced Craft Certificate in Carpentry and Joinery. The members of this Institute are judged to be extremely competent in their craftwork.

The Institute of Plumbing (IOP)

This Institute was formed in 1906 and then merged with the Registered Plumbers Associated in 1970. Usually, members are practising Plumbers and Plumbing Contractors. For Associate Membership (AMIOP) an appropriate qualification in Plumbing is necessary. For Corporate membership grade, that is, Fellow and Member, the minimum age is 25 years before the designated letters of FIOP and MIOP can be used.

Guild of Bricklayers

The Guild's aim is to promote high standards of work within the bricklaying trade. In order to do this members and their guests are invited along to functions normally with an educational/training theme. Films, lectures and visits are the more usual ways of presenting new ideas and methods to members. There are also social functions. Members have usually gained an Advanced Craft Certificate in the subject of brickwork.

Plasterers' Craft Guild

This is a small Guild but gives a worthwhile service to the Industry. Because of the interest displayed by its members in organising meetings and competitions, the standards in plastering, which have fallen over the years, are being maintained at a level which is high enough to meet the challenges which are made upon it. Renovation work to the country's noble, historical buildings particularly calls for a high degree of plastering craftwork, and some designers still call upon Plasterers to produce a limited amount of high-class work.

Other guilds

There are in London many Livery Companies which are either Trade Guilds or Craft Guilds. The members of the Guilds help to promote standards of training for young men in their represented trade/craft, and help to encourage school-leavers to train in the trade/craft. Such Livery Companies are the Masons' Guild, Glaziers' Guild, Carpenters' Guild and Stainers' and Painters' Guild (see Fig. 2.2.2).

2.2 Historical development

After the Norman conquest in A.D. 1066 many stone castles and fortifications were built to protect the Barons from possible uprisings by the British. In some cases these castles were constructed in the centre of the towns or villages; usually on points overlooking the area. In other cases, a site was selected in a position which could be easily defended but was some distance from habitable districts. Very soon villages began to develop close to the castle for protection against intruders from other areas. Church building followed where new villages or towns grew, and sometimes these expanded into monasteries.

Large numbers of craftsmen were required continuously for such projects as church and castle building, the most common being the mason and the carpenter. Castles had to be constructed quickly, so usually a large labour force of masons was housed in Lodges until such times as the work was completed. With the church building programme a longer time could be afforded and the Lodges which were set up became permanent, particularly where monasteries were concerned. These structures became conglomerates, with all types of additional buildings for storage of food, hospitals, sleeping quarters, libraries, schools and many more. The land adjoining these ecclesiastical buildings was used for agriculture which made the whole monastic unit self sufficient.

The Lodges were bases for the masons in particular, who became very knowledgeable at their craft and jealously guarded their skills. Mystery surrounded their craft. They would only expose their knowledge to the young apprentices who came under their wing for training. Such was the skill of these people that seven-year apprenticeships were thought necessary to graduate into craftsmen, sometimes referred to as journeymen because later they had to be willing to journey round the estates or from one area to another in search of work.

In the later period of the fourteenth century, work was plentiful in some areas because of the recurring plagues which reduced the population. There arose the problem of too few craftsmen chasing too many jobs. Wages rose because of the demand for labour, and there developed the new journeyman. In some parts of the country there was a drifting of labour into the towns, but many of the craftsmen already living there formed themselves into Guilds (Gilds) to protect their livelihood.

Craft Guilds developed rapidly especially when work became scarce, and usually only those residing and who were trained in the towns could practise their craft, which eventually restricted the movement of labour and put Guild members in a very privileged position.

The Craft Guilds became very influential and many of the members held important positions in the towns, such as Mayor and Eldermen (Aldermen). There were two levels of membership of the Guilds, the masters and journeymen. The apprentices gained admission after seven years' training. The masters were those who now had their own businesses and who employed some journeymen or apprentices; the journeymen having served apprenticeships but were unable to acquire businesses because of insufficient capital, and the apprentices were usually bound to the master, sometimes with little reward, until they passed out of their time and became journeymen.

Shops were the normal places of work for the craftsmen, and today they are still referred to as Workshops. One could go to the shops to purchase certain wares and at the same time observe there the craftsmen and apprentices working. Naturally, many of the construction workers' places of work was on the construction sites. The masters became rich and soon acted as merchants, leaving the craftwork to the journeymen. They then took control of the guilds and began to exclude the journeymen by making new stringent rules and by increasing subscriptions. Later, induction banquets had to be given by new members and only the richer journeymen could afford this, the sons of existing masters or merchants being usually the only ones gaining admission. The journeymen then had to set up their own craft guild, whereas the merchants (as the masters were then soon called in many craft guilds) had control of what developed into the trade guilds. Many of the craft and trade guilds were formed into Livery Companies by Royal Charter in the City of London. This gave some of the liverymen the right to trade nationally instead of just locally.

In the building industry the master craftsmen became employers of labour, especially the master masons and master carpenters. These masters were normally approached to undertake work similar to the modern day contractors, and they used journeymen to actually do the construction work while they supervised the project. Design work was often done by these masters until a new breed of journeyman/master took on the role of principal master; the earliest form of architect in this country. This evolved about the fifteenth century, and by the seventeenth century architectural designing was well established, culminating in the eighteenth century to persons actually receiving formal training in offices of skilled Architects which had never been thought necessary until then (see Fig. 2.2.1).

Meanwhile, the craft guilds began to organise themselves into Friendly Societies providing social benefits for their members in the way of sickness benefits, unemployment benefits and benefits for deceased members' widows. The Industrial Revolution brought large workforces together in the cotton and engineering industries, which led to the workers combining first into Local Associations, and then into National Unions later. This was maintained by the skilled workers in the nineteenth century, but by the 1880s unskilled workers also formed themselves into unions. With the amalgamations which have taken place throughout the country since the turn of the twentieth century some unions are many millions strong in membership (see Fig. 2.2.2).

The Industrial Revolution assisted in influencing people to try new skills and from it developed, firstly, the civil engineer, then the quantity surveyor, next came the structural engineer and the numerous services engineers (see Fig. 2.2.1).

The City and Guilds of London Institute

This examining body was founded in 1878 by the City Livery Companies and the City Corporation. It was designed to raise the standards of technical efficiency through training and education to meet the increasing challenge from abroad.

In 1900 it was granted a Royal Charter. The Institute is an independent body although contributions can be received from the Corporation and the Livery Companies.

The work of the ancient craft guilds is carried on through the Institute, and now both craftsmen and technicians' examination courses are offered.

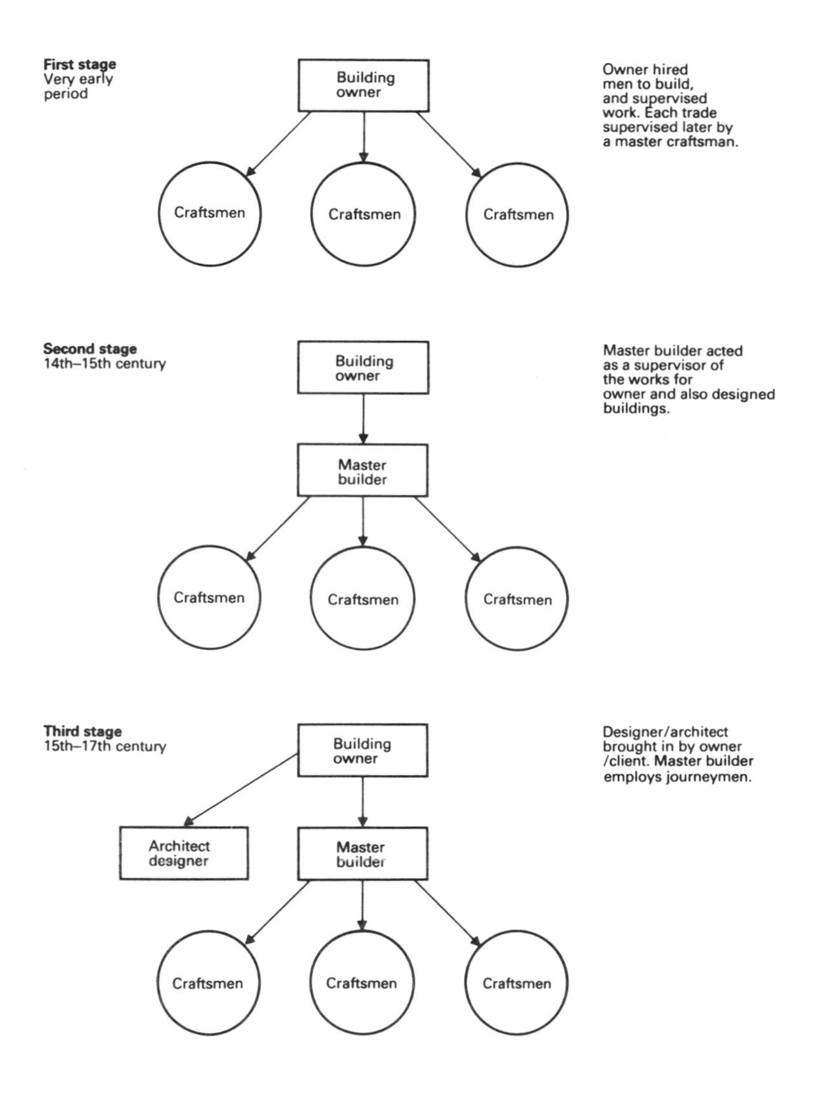

Fig. 2.2.1 Evolution of the building industry

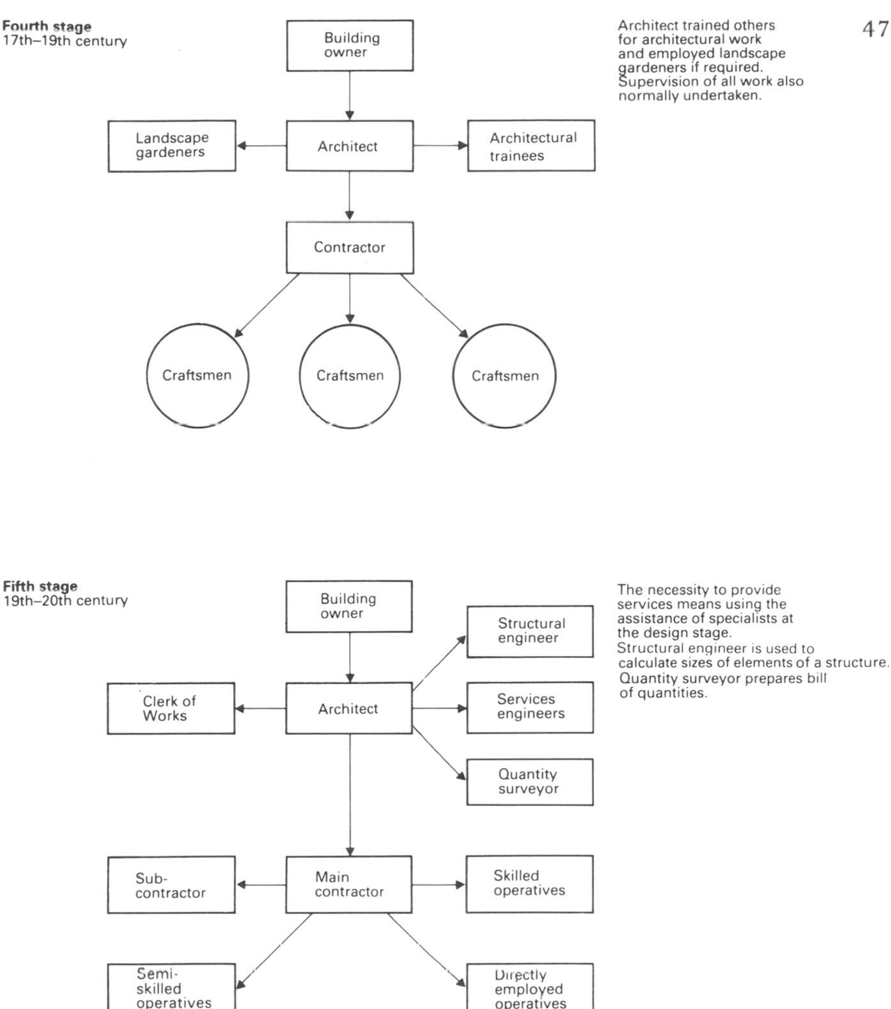

Fig. 2.2.1 (continued)

Control of the Institute is almost entirely vested in its council, i.e.:

1. *Ex-officio* members – Lord Mayor and presidents of certain learned bodies (societies and institutes), and representatives appointed by the City Corporation.
2. Representatives from the subscriber City Companies (Livery Companies).
3. Members who are voted on to the Council who can greatly contribute to the Institute.

7. The Joinery and Timber Construction Association.
8. The Association of National Stone Industries.

The Federation of Civil Engineering Contractors

The organisation was formed in 1919, and is the employers' organisation for the Civil Engineering Industry. The Federation is the representative body for Civil Engineering regarding all aspects which affect its membership of approximately 600. It has a system of standing committees which are responsible to the main Council of the Federation. The Council governs the Organisation and makes decisions with the interest of the members in mind, after consideration of the national policy of the Federation. Separate standing committees meet to discuss and later report back to the Council on the following matters:

1. Wages.
2. Condition of Contract.
3. Safety, Health and Welfare.
4. Common Market.
5. Research.
6. Training.
7. Daywork Schedule.
8. Future Planning.

The Federation soon established a national machinery for collective bargaining with the employees' trade unions, through the meetings held with the unions' representatives, which later resulted in the formation of the Civil Engineering Construction Conciliation Board for Great Britain.

All parties to the Conciliation Board meet regularly to discuss wage claims, working conditions and disputes which cannot be settled at a local level.

Due to the Federation, Civil Engineering Contractors have a better 'Condition of Contract' after an agreement was reached with the professional bodies of the Institute of Civil Engineers and the Association of Consulting Engineers, in 1950.

Consultations take place with the Government regarding existing and proposed legislation, and at all times close ties are kept with the Institutions of Civil and Municipal Engineers, Association of Consulting Engineers and the Society of Civil Engineering Technicians regarding matters of mutual interest.

Association of Consulting Engineers

Employers in professional practice in engineering usually join this Association to safeguard their interests. The Association will advise the public of the standing of its members in the profession regarding qualifications and experience, if so asked. The usual requirements for membership being that members should be qualified and be full members of a senior Engineering Institute, such as, the ICE, IStructE, IEE or IMechE, and be in a Consultancy Practice.

Other employers' organisations

There are many federations/associations of specialist firms operating within Great Britain, but too many to mention in this volume.

The following is a list which is in no way exhaustive or arranged in order of importance:

Electrical Contractors' Association.
Heating and Ventilation Contractors' Association.
Federation of Piling Specialists.
Federation of Associations of Specialists and Subcontractors.
National Federation of Plumbers and Domestic Heating Engineers.
National Federation of Roofing Contractors.

2.4 Employees' organisations

The major representative union of employees in the construction industry is the Union of Construction, Allied Trades and Technicians (UCATT). This Union, like so many others, helps to safeguard the interests of its members regarding their safety, health and welfare, and negotiates with the employers on matters such as: wages and the training of operatives. There are, in addition to UCATT, other unions which have a special interest in the Industry, although only a small percentage of their members are directly employed within it. A large proportion of the construction labourers and plant operators are members of the Transport and General Workers Union (TGWU). Some craftsmen also belong to TGWU. Naturally, the respective sections of the TGWU have close ties with UCATT. The other unions are: the General, Municipal and Boilermakers Union (GMBU) and the Furniture, Timber and Allied Trades Union (FTATU). The former represents local Government workers in construction and public utility services organisations (gas, electricity, water) and boilermakers, the latter embodies those working as wood machinists and furniture trade craftsmen (see Fig. 2.4.1 for the UCATT organisation).

There are other employees within the Industry which have unions to represent their interests. Mechanics, fitters and welders are workers who form the support section of the Industry. They are a very important group, and employers may be called upon to negotiate with them periodically when disputes arise. Plumbers and electricians and other electrical craftsmen have combined to form the Electrical, Electronics and Telecommunication Union/ Plumbing Trades Union. This union tends to negotiate separately with employers.

Both in the building and the civil engineering industries UCATT and the other construction employees' unions belong to the NJCBI and the Civil Engineering Construction Conciliation Board (CECCB).

Each union is a member of the Trades Union Congress (TUC), which acts in a similar capacity for the employees' unions as the Confederation of British Industry (CBI) does for the employers' organisations.

Recently, due to dissatisfaction with the arrangements under the NJCBI and their lack of consultation powers, the members of the Federation of Master Builders (FMB) together with the TGWU (Construction, etc.) have formed a joint body representing smaller builders and the TGWU.

2.5 Trade associations

With the production of so many varied materials which are used within the construction and other industries, there has developed over the years numerous Trade Associations primarily concerned with research into the use of the materials to which they are associated. These Associations were set up by the producers and manufacturers of the various materials from home and abroad. It is obvious that those organisations associated with the production and use of lead, for example, would combine to finance an association such as the Lead Development Association. The responsibility of the LDA is: to develop the use of lead; to give advice to its members; and to educate the public regarding lead products.

There are many Associations which look after the interests of its members and their work is briefly described below:

Brick Development Association (BDA)

The membership of this Association manufacture more than 80 per cent of the total output of clay and sand and lime bricks, and therefore plays a major role in promoting the use of bricks throughout Great Britain. In furthering this product, it publishes Technical Notes and the Brick Bulletin at regular intervals. Education films are produced and research projects may be sponsored.

Cement and Concrete Association (C & CA)

Most Associations are non-profit making, the C & CA being no exception. The use of cement and concrete is encouraged particularly in the building, structural and civil engineering fields. The C & CA, therefore, publishes literature suitable for distribution to the various users of the products, which advises on techniques which should be adopted to obtain the best results.

There is usually a very varied programme of courses held by the C & CA at its Training Centre. The courses deal with all aspects of concrete design, practice and principles.

There are numerous publications by this Association – the *Concrete Quarterly*, and *Magazine of Concrete Research* being but two of them. Films and slides are also available on request for lecturers and training officers.

Copper Development Association

This Association, like most others, is a non-trading and non-profit making organisation which is financed mainly by copper developers from all over the world. A free Technical Service is available where queries can be answered

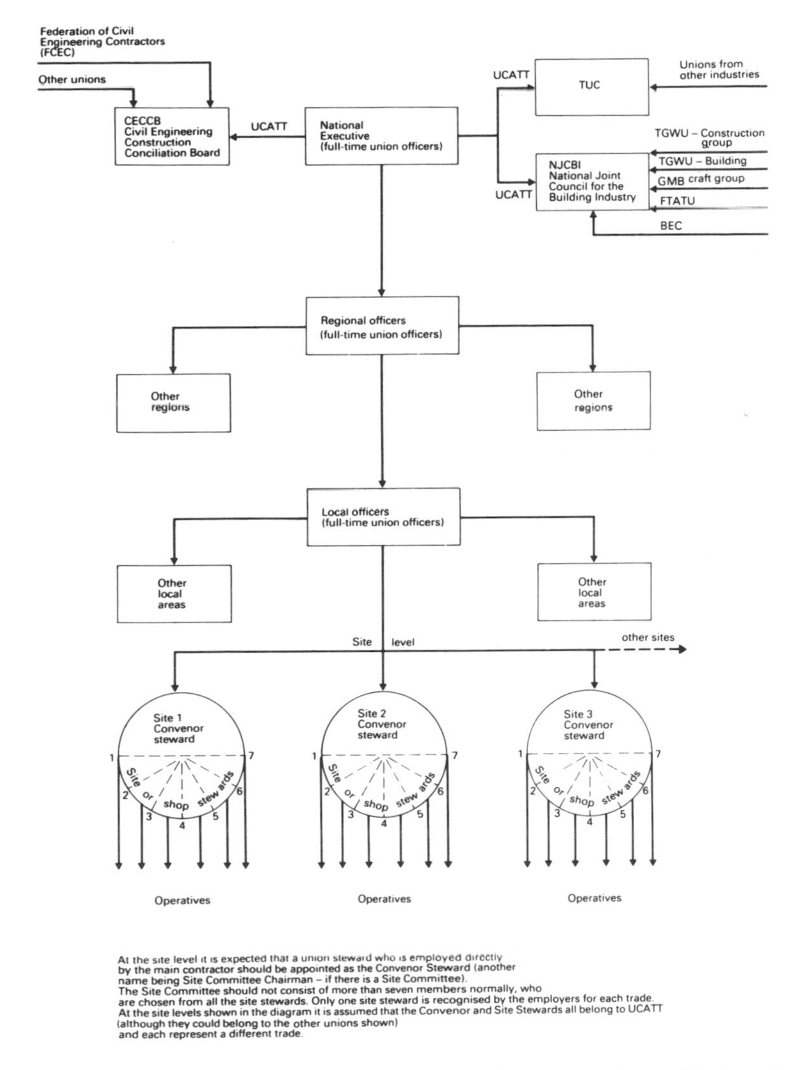

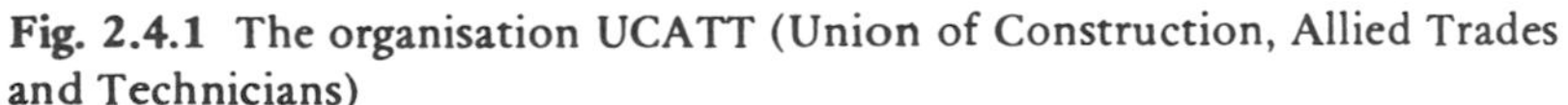
At the site level it is expected that a union steward who is employed directly by the main contractor should be appointed as the Convenor Steward (another name being Site Committee Chairman – if there is a Site Committee).
The Site Committee should not consist of more than seven members normally, who are chosen from all the site stewards. Only one site steward is recognised by the employers for each trade.
At the site levels shown in the diagram it is assumed that the Convenor and Site Stewards all belong to UCATT (although they could belong to the other unions shown) and each represent a different trade.

Fig. 2.4.1 The organisation UCATT (Union of Construction, Allied Trades and Technicians)

over the telephone or by letter. The publication, *Copper*, is available periodically along with other research information, and like most other associations free films are made available to promote the use of copper and its by-products.

Producers continuously search for new areas to develop the use of their materials and competition is very keen, particularly in the field of plumbing and building services. Lead was used extensively for pipework in the early part of the twentieth century for water supply within a building, and was then followed by copper after the Second World War. Now, because of the expense of copper, stainless steel and, in particular, plastic pipework also shares this corner of the market. Naturally, each association tries its best to enhance the properties and advantages to the public of the use of its sponsors' products.

The following are but a few of the other Trade Associations:

Aluminium Federation.
British Plastics Federation.
Flat Glass Association.
Gypsum Products Development Association.
British Wood Preserving Association.

2.6 Other organisations

There are organisations attached to the construction industry which play a major role not only in maintaining standards, but in conducting research, inspecting standards of workmanship on-site and contributing generally in providing for the needs of the Industry. Some of these organisations were created by the Government and the others by professional or trade organisations. The more prominent of these bodies are dealt with here:

Construction Industry Training Board (CITB)

As in most industries this Industrial Training Board was formed in 1964 after the passing of legislation by the Government. It was to promote all aspects of training for the Construction Industry, and in so doing, was to spread the cost for the training over as many firms as possible. It is now one of the few training boards still in existence after the MSC came into existence in the 1980s.

The Board is empowered to raise a levy from the Industry so that grants, for all training purposes, may be made to firms who allow their employees to undertake training which is acceptable to the Board.

Levy: The levy varies from year to year depending on the needs of the Industry and is a charge made against a firm's annual payroll. The small firms with a low payroll are exempt from the levy, but they cannot claim a grant if their employees attend training courses or in-service training schemes.

Grants: These are fixed yearly and are payable for each apprentice, trainee, articled pupil, and others who receive training approved by the CITB. Firms must claim back the money paid previously to the Board through the levy after sending employees on apprenticeship training or courses at Colleges of Technology. In-service training by the firm is also suitable for a grant. Management trainees are also eligible for grants if sent on recognised courses.

Training Centre: This is situated at Bircham Newton, Norfolk. There are other Centres which are used by the CITB. The types of special training undertaken at the Centre by trainees covers: earthmoving plant, cranes, road work equipment, steelfixing, scaffolding, concreting and many more – semi-skilled and skilled work. Supervisory courses are also organised at the Centre.

The Board: This comprises of a Chairman and Deputy, employers' representative from building, civil engineering, electrical, heating and ventilating subcontractors' firms, operatives' representatives and representatives from Government departments and education.

The Board has general responsibilities for administration and policy making and is assisted by Committees in building, civil engineering, electrical engineering and mechanical engineering.

The National House Builders' Council (NHBC)

In 1935, House Builders were becoming concerned at the bad image they had in the eyes of the public so to encourage good standards of house building and to expose the unscrupulous individual or firm who posed as reputable builders, the NHBC was formed. Since its formation most *bona fide* house builders gained membership, the odd disreputable builder gained membership at first, and then, due to complaints made to the Council by clients, have been exposed as unacceptable members of the NHBC.

Membership: Normally, new private house builders become probationers until they have completed a number of houses to the required standard. They would then enjoy full membership provided they are financially sound.

Fees: There is a registration fee and a further fee payable for every house inspected, depending on its value, which is allocated a NHBC Certificate of satisfactory standard. The inspector has the right to access of any housing site of the NHBC members in order that the standards laid down are enforced. Certificates will not be issued for any house which is sub-standard.

Advantages of membership

1. Many Building Societies only allow mortgages for new dwellings which carry an NHBC Certificate.
2. NHBC members enjoy the goodwill of the public because this independent watchdog helps to maintain high building standards (members are allowed to use the Insignia of the NHBC, which is a lighthouse, on their letter-headed paper).
3. Members are protected, by insurance, from defects which may develop after work has been completed. The client can then claim through the NHBC instead of going through the expensive process of suing in a Court of Law.
4. A builder who has unreasonable demands made upon him by the client may call upon the NHBC to give an unbiased opinion on the problem, thereby, in some cases, protecting himself from harassment.

Property Service Agency (PSA)

This Agency is a section incorporated within the Department of the Environment (DOE), which was set up in 1970 by the Government. The DOE covers the work of three previous separate Ministries: Ministry of Housing and Local Government, Ministry of Transport and Ministry of Public Building and Works.

The PSA helps to satisfy the needs of other Government departments including the Armed Services and public sector organisations such as the Post Office, both at home and abroad. Its interests lie in land owned by the Government, and accommodation, buildings, maintenance and installations. There are employed within the PSA the following classes of employee: executive, professional, technical and clerical officers, tradesmen and unskilled operatives.

The work undertaken by these Civil Servants is wide and varied. Those who are directly associated with output in building and civil engineering are: architects, civil engineers, structural engineers, quantity surveyors and services engineers, including their associated technicians.

The size and complexity of the Agency's construction work ranges from dwelling houses, maintenance, building extensions, dockyards, airfields and power stations.

Concrete Society (CS)

This organisation was created to help in the development relating specifically to the production of concrete and associated works. It encourages research, training and education in concrete work, and produces booklets, journals and papers regularly on topics such as reinforced concrete and formwork design. It liaises with the Cement and Concrete Association and the Institute of Structural Engineers on topics of mutual interest.

Members of the CS are situated in all sections of the industry. They include: civil and structural engineers, civil engineering contractors, local authorities, architects and the PSA.

Building Advisory Service (BAS)

The BEC (Building Employers Confederation) was responsible for setting up BAS, which they founded in 1954. Within this organisation there operates four divisions, each of which is responsible for providing a service to the building industry generally.

BAS is financed mainly by the many firms which uses its services. The divisions are:

1. Safety Division.
2. Consultancy Division.
3. Appointments Division (BASAD).
4. Management Training Division.

While discussing the BEC it is as well to mention yet another organisation which comes under its wings. As security is causing more and more concern within the industry, a self-financing organisation called CONSEC (construction security) now operates on a consultancy basis, giving advice to firms on security systems and measures to prevent the unacceptable level of theft, pilfering and robbery.

1. Safety Division

This operates on a consultancy basis as well as being a provider of advice and information to those who seek it. Safety courses are held regularly and assistance is given on the safety content in management courses or seminars. Whenever new legislation and codes of practice are introduced on safety one can be sure that the Safety Division will have played its part in representing the Building Industry. Mobile Safety units operate throughout the country visiting sites where requested to do so by firms who wish to give their operatives some instructions on safety.

Finally, there is available the very popular Construction Safety Manual which can be purchased by any firm, to be used by their safety officers or managers.

2. Consultancy Division

This division has many clients throughout the country in the form of builders, government departments, civil engineers and many more, and provides assistance on such management techniques as: planning, programming, incentive schemes, marketing and most problems encountered by management.

3. Appointments Division

Recruitment of technical and management staff is one of the specialities of this division. It has expert staff who take over, amongst other things, the advertising and selection process from firms, and eventually submit short lists of candidates.

4. Management Training Division

Courses, conferences and seminars are held on almost any subject relating to management. In-company courses are also held when required, which can be designed to suit any companys' needs. Advice will also be given on company training programmes.

2.7 Research and development

If the Industry is to continue to cope with the demands made upon it by the legislators, conservationists and society generally it must invest in research in order to develop its resources to the full. There is a need to keep costs to the minimum and to help conserve vital energy which adds to the cost of Britain's already stretched economy.

The investment into research helps to bring about changes so that improvements can be made to building types, design, construction techniques and building materials.

Research is undertaken by many organisations, the Government being one of the biggest financial contributors. Since it was recognised in 1921, when the Building Research Station was set up – the first of its kind in the world – that building research was essential, new innovations have been numerous particularly within the last twenty years.

The introduction of lightweight concrete, internal wall and ceiling finishes – such as drylining instead of wet plaster – system building and management techniques in the form of Critical Path Networks, are but a small percentage of new ideas brought about by the many different organisations who carry out research yearly. Development has been rapid and still continues at an ever increasing rate.

The new innovations have greatly enhanced the contractors' position and have contributed to making the Industry even more efficient than it was.

Research organisations

1. *Building Research Establishment (BRE)*

This combines the work of the following organisations:

(*a*) The Building Research Station (BRS) and Building Research and Advisory Service.
(*b*) The Fire Research Station (FRS).
(*c*) The Princes Risborough Laboratories (Forestry Products Research).

Building Research Station: The Government finances the work undertaken at the Station, which is situated near Watford. The type of research and other services offered to the industry are as follows:

1. Materials and construction research.
2. Geotechnics research.
3. Physical requirements and engineering services research.
4. Building operation economics and mechanisation research.
5. Information activities – distribution of Research Digests.
6. Advises on the preparation of: BS, BS Codes of Practice, and Building Regulations.
7. Education.

Regarding the other two research centres, the Fire Research Station tests structures and materials to destruction, and advises on new standards; the Princes Risborough Laboratories researches into the efficient use of timber and timber science generally, such as: the seasoning of timber, preservatives, machinery, and man-made products (plywood and other boards).

2. *Construction Industry Research and Information Association (CIRIA) – 1967*

This is a body which helps to finance research in the various fields of construction and may allocate funds to universities, industrial firms and other research organisations for special project research.

3. Large building and civil engineering firms

Many large contractors carry out research and the results are usually patented and are used by their own parent companies, unless they agree to license its use to others. One such organisation is John Laing Research and Development Ltd.

Apart from the testing laboratories which are used for the testing of materials and other structural components, the research sections hold courses for supervisors on the design and performances of materials and plant.

Soil mechanics generally forms a large part of some contractors' research programmes. Methods of stabilising soils is also dealt with, particularly when so much damage is caused to structures through settlement due to the sub-soil compressing unnecessarily.

4. Independent testing laboratories

These exist throughout the country and usually are Quality Control Research Organisations who issue test certificates and reports on concrete cubes and other routine matters. Contractors who wish to get independent results on materials produced on-site seek the services offered by these organisations and in exchange pay a fee.

The other services offered by these laboratories are: soil testing, materials testing and advice on materials and other construction failures.

5. British Board of Agrément

This is a Government-sponsored organisation and is partially financed by the manufacturers of new products.

Architects dislike using new products unless they have proof of the products' capabilities. The manufacturers, therefore, obtain an independent test certificate by approaching the Agrément Board who will then put the products to test. If the results are satisfactory a certificate is issued, and can then be used by the manufacturer to substantiate the claims made of the product. The certificates are widely recognised and can be used until such times as a British Standard is prepared.

6. Timber Research and Development Association (TRADA)

The important point regarding this organisation is that it works independently from producers and manufacturers of the raw material. However, it is financed by the Government and members' contributions. The members can be individuals or firms who require to know the latest developments and suggestions put forward by the association.

The type of research and services offered by TRADA are as follows:

(*a*) Research into the efficient and economical use of wood for all industries, i.e. architecture, construction; boat building and anywhere that requires the use of wood.
(*b*) Library service for reference purposes.
(*c*) Advisory service and information bulletins are available on request.
(*d*) Education service – films, teaching aids, and other technical sheets.

7. Universities, polytechnics and colleges of technology

The courses which are associated with building and engineering have incorporated within them scientific aspects of study. The undergraduate/post graduate/student in many cases carries out research while undertaking project work. The results usually obtained during these studies are often published to contribute to the developments which take place within the construction industry. However, money is needed to finance some of the more spectacular work and as a result firms allocate grants to these bodies, provided that any new developments can be made available to the sponsor firm.

The universities have been known to act as independent bodies and the professors have taken on the responsibility for the research, with the assistance of a few postgraduates, where disputes have arisen between the designer and contractor when there has been a construction failure. The questions asked usually are: 'Was it a design fault?' or, 'Was it poor workmanship?' As an independent body they can research into the problem and give an unbiased opinion or result.

A few colleges give extra services to industry by allowing contractors and others access to their construction libraries, which are usually based on the CI/SfB System. They will also advise builders, if they require it, on construction methods, administration problems, industrial training and education. Material testing facilities are usually available but a charge is made for this.

8. Other research organisations

When a raw material is mined, quarried or processed, the producer requires an outlet for his/her products. Popularity of the various materials vary each decade and, consequently, one material is preferred to another for the production of, say, water cisterns in roof spaces. At one time the cisterns were made using lead; then cast iron; later galvanised steel became popular. The reasons for this and changes elsewhere is due to the ever changing costs of the raw materials, and also the efforts of the trade associations which carry out research to find new ways of using the materials which will add to their popularity in use.

Trade associations are usually financed by the producers of the raw materials and manufacturers. It will be these same members of the trade associations who benefit ultimately from the new ideas brought about by the research undertaken by the association, many of which have been previously mentioned. There are other organisations such as:

Constructional Steel Research and Development Organisation.
North Wales Slate Quarries Association.
Paint Research Association.

It must be noted that there is a different trade association for nearly all the producers of a wide range of materials produced for the construction industry.

Information regarding new ideas and products need to permeate through to the Industry if architects, contractors, and clients are to benefit fully from the new discoveries. Suitable methods of communication should be available for the distribution of information, and the Industry at present has the following, which appear to work quite efficiently:

Her Majesty's Stationery Offices (HMSO): Any legislative instruments and other Government publications can be purchased from these offices, and it is the major source of issue of printed Government material.

Building centres: There are a number of these Centres situated in the larger towns or cities of Great Britain. They are display shops for the manufacturers of new products. These products, including the associated literature, are shown on the display stands situated round the shops and for this service the manufacturers are charged a fee.

Trade associations: These organisations are only too pleased to send, on request, their publications regarding each member's products.

Professional and other institutes' journals: These tend to be very important documents for the dissemination of information, either into research undertaken by trade organisation or other research bodies. There are, however, many individuals associated within construction who conduct their own research and allow the results to be published. They are a valuable asset to all construction organisations and societies, and contribute to the Industry's development.

Members of learned bodies are encouraged to produce white papers and written articles which may be used in their monthly journals. Such bodies are: the Royal Institute of British Architects through its *Architects' Journal* (*AJ*), the Institute of Building Journal, called *Building Technology and Management*. Naturally there are many more institutes and each produce a journal on a monthly basis.

Barbour Index Limited: This organisation provides a service to those firms which have a construction library within their premises. A fee is charged for the service, which includes the provision of a back-up enquiry service as well as a specially arranged library to facilitate the up-dating and sorting of information. A representative from Barbour Index Limited visits each firm's library monthly, and incorporates into the system manufacturers' new literature and any other literature relevant to construction.

There are other organisations similar to Barbour Index Limited which operate on a business footing – the Building Products Library Service being but one. The system for filing used by these organisations is that which is administered by the RIBA called, the CI/SfB System.

9. *British Standards Institution*

This organisation was set up as a non-profit-making organisation and under Royal Charter in 1929. It attempts to coordinate the efforts of all manufacturers and persons associated with the use of products. It lays down standards for the improvement, standardisation and simplification of all materials. It also lays down standards of quality and dimensions. Representatives on the BSI are nominated by the founder organisations, i.e. TUC, CBI, Nationalised Industries, Ministry of Defence, the DOE, etc.

There are five kinds of documents produced by the BSI which are:

1. BS (British Standards).
2. CP (Codes of Practice).
3. DD (Draft Development).
4. PD (Published Documents).
5. DC (Drafts for Comment).

British Standards were initially product specifications, but now they include schedules, methods of testing, basic data, etc. The Standards are laid down to ensure that quality, performance and usability of components are met, and also lays down preferred forms of a product and properties of a finished article, and method of testing for verifying that the standards have been achieved.

There are BS for all kinds of products, such as: tools, plant, equipment, drawing practice, design, and hundreds more.

Codes of Practice are recommendations for good practice to be followed during design, manufacture, construction installation and maintenance (workmanship) with a view to safety, quality, economy and fitness for purpose.

Draft Development documents are used before British Standards are issued where firm standards cannot be stated because of the lack of information on the introduction of a new idea or subject. The documents (DDs) are intended to be used for a limited time only until feedback due to experience, knowledge and usage is accumulated to allow a DD to be converted to a BS.

Published Documents (PDs) are for publications which cannot fit into the other group categories, but are seen as suitable until sufficient information and experience feedback allows a BS to be issued on a topic.

Drafts for Comment (DCs) are issued once the BSI is convinced that a subject is important enough to provide resources to look into the preparation of a BS. Anyone can request a new standard. The Institution may prepare a number of revised drafts of a project for issue to the public. Anyone who then has an interest and wishes to express an opinion on any proposals may do so and, hopefully, after many ideas are considered and suitable period has elapsed a BS may then be introduced.

The BSI has a registered certificate trade mark (a Kitemark) – see Fig. 2.7.1. It is an assurance that products have been produced by a manufacturer under a system of supervision, control and testing during manufacture, which includes periodical inspection by inspectors from the BSI. A licence must be paid for use of the Kitemark by material producers.

Fig. 2.7.1 BSI Kitemark.

Chapter 3

Clients and the building construction team

3.1 Types of clients and market sources

It is the client who enables a firm or company to stay in business, and therefore those associated with the output of the industry rely on the way their firm's or company's good name is maintained. To this end the management, supervisors and operatives all have responsibilities to create 'goodwill' or uphold the good name the company has in the eyes of the public and prospective clients. This does not only apply to the contractor, but equally to the architect, quantity surveyor, engineers and other members of the design team. In particular the architect must endeavour to create 'goodwill' in order to obtain 'commissions' from clients in whose service each member of the construction industry rely.

While there are many ways in which architects seek an outlet for their professional services each may find a particular one which recurs more often than others. The usual methods of obtaining commissions however are described as follows:

1. By competing with other architects in design competitions initiated by clients and their representatives.
2. Through personal contacts with friends, relatives and acquaintances.
3. Continuous contact with the public through participating in outside interests.

4. Recommendation, usually by past satisfied clients who have received a good, professional service. Even contractors and fellow members of the team could give a complimentary recommendation to their friends and clients.
5. By invitation through their professional status in the community.
6. By continuous employment as consultant by standing arrangements with a client, say, church commissions or factory owners, and other owners of property.
7. The RIBA will, if asked to do so by a prospective client requiring architectural services, issue a list of architects' names who are able to undertake work which the client has in mind.
8. By advertising under the strict code laid down by the RIBA or other similar bodies.

When an architect is appointed by the client confirmation is necessary in writing to the client to make it legally binding, and to make clear the conditions by which an architect is normally engaged. Clients range from private individuals to consortiums of local authorities, and while architects are employed in central government bodies and local authority offices, many projects are designed by private practice architects particularly when prestige buildings are contemplated.

Contractors on the other hand can obtain work by methods which, in some cases, requires an energetic approach by themselves or their representatives.

They should be constantly seeking ways to show themselves favourably in the eyes of the public, and, in their efforts to maintain full order-books, should be vigilant to the many systems clients adopt to offer work to contractors. To these ends the contractor could possibly do the following:

1. Make personal contacts – with those who are friends, relations or acquaintances of the company.
2. Advertise services: in local or national newspapers; in trade periodicals or journals; on radio or television; on bill posters; in telephone directory Yellow Pages.
3. Design suitable trade marks or company markings on vehicles and plant.
4. Satisfactorily design signboards: for head office; for entrances to contract sites.
5. Issue calling cards to company representatives for distribution.
6. Canvassing: personal approach or by letters and leaflets.
7. Gimmicks: competitions in papers to encourage public participation and, therefore, create interest in the company's activities.
8. Public relations exercise: laying of foundation stone; progress report on a project in local or national papers; topping out ceremony; handing over ceremony.
9. Forming a consortium: for special work, i.e. very large or specialist contracts.
10. By speculation: purchasing of property to develop for sale.
11. Inclusion on local authority list: rotation method.

12. Creating goodwill: satisfying the clients and architects by completing projects to a good standard, and on target.
13. Replying to advertisements where there is an 'open tender' situation.
14. Energetic attendance of local meetings in trade/commercial, professional and public.
15. Show examples in efficiency: provide public viewing platforms at sites; neat layout of site huts and materials; provide good welfare facilities which may encourage operatives to commend the company.
16. Smart appearance of all property belonging to the company.
17. Promotion parties: invited guests to include past and present clients and important dignitaries to see new line of business offered or improvements made to the organisation or company buildings.
18. Marketing (the term used to denote the previous 17 activities).

This list by no means exhausts the action which contractors can take to promote the wellbeing of the company, but does go some way to help it to continue functioning satisfactorily.

Clients from both the public and private sectors prefer to use the services of an overall manager to design and supervise their projects. These managers are usually either architects, structural engineers or civil engineers. They can, of course, expect from some large contractors an all-in service or 'package deal', where the contractor uses his own team to design and construct the entire project. Contractors also offer a Management Fee system – they manage the project and use the construction services of other contractors as subcontractors.

All those of the construction industry, whether they are in professional private practice or operate as nationally operating contractors, look to the following clients for business:

1. Central Government: through the various Ministries.
2. Special Organisations: Airport and Port Authorities; British Aerospace Industry; British and other Petroleum Companies; British Telecom; British Rail; Central Electricity Generating Board; National Coal Board; British Steel Corporation, etc.
3. Special Public Corporations: British Broadcasting Corporation; New Town Development Corporations; London Transport; Metropolitan Water Board, etc.
4. Housing Associations or Trusts.
5. Companies and Groups of Companies.
6. Partnerships.
7. Local Authorities – housing, highways, public health, schools, etc.
8. Sole traders.
9. Private individuals.
10. National and other Trusts.
11. Foreign Governments – Embassies and overseas work.

3.2 Architects, specialists and other members of the design team

Architects in private practice operate either as sole traders or partnerships and adhere to very strict codes of professional procedures. Designers or design service agencies' businesses are created to undertake similar work but in most instances safeguard their position by operating as limited companies. Architects rely almost entirely on the reputations they gain through the impacts

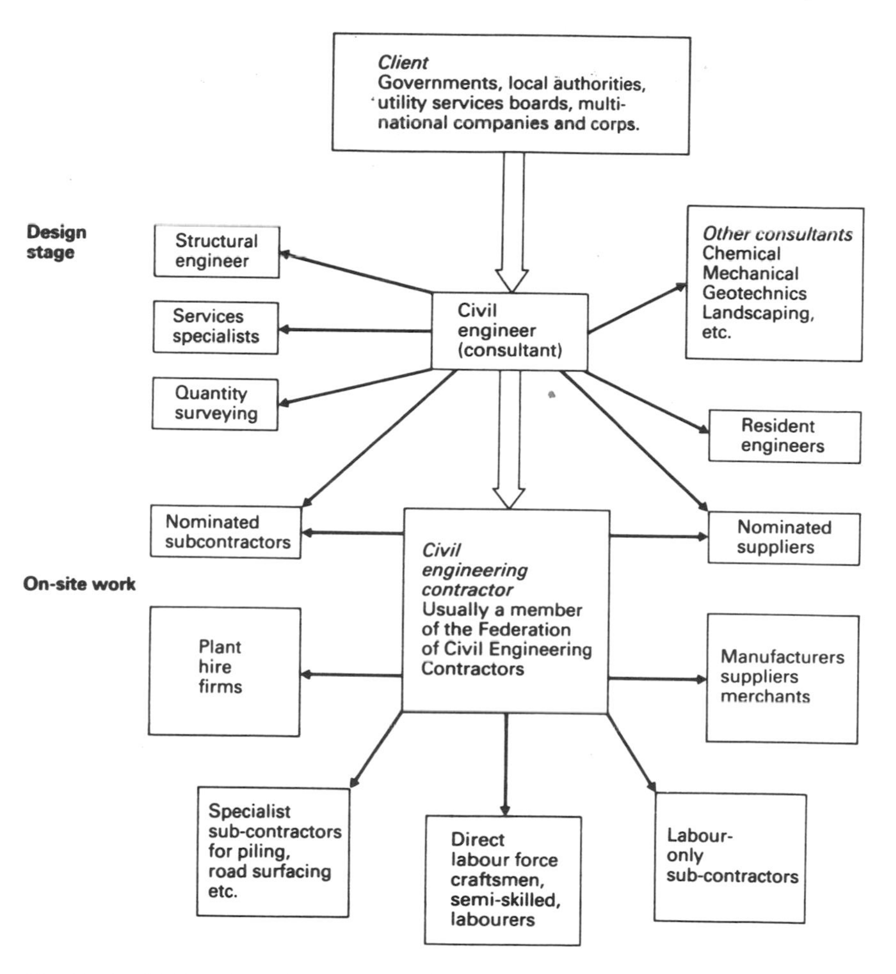

Fig. 3.2.1 Civil engineering structure (present day)
Note: The consultant civil engineer acts in a similar capacity to an architect and designs many structures such as: roads, bridges, canals, docks, railways and other major works. He may still require the assistance in his design work of the consultants and specialists as shown above. His supervisor on-site is the resident engineer

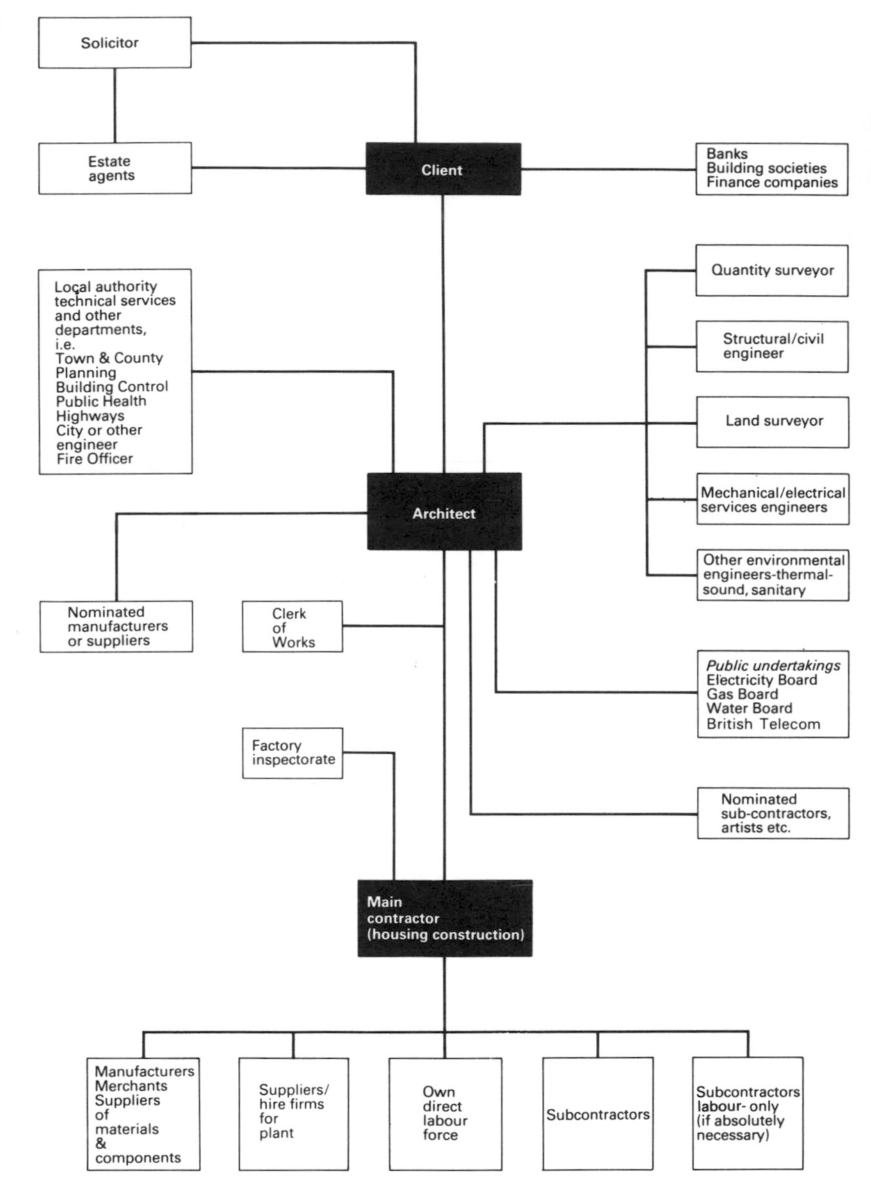

While the architect appears to employ all designers, consultants, contractors, etc. He does so in this example by agreement with the client. Usually the architect recommends the appointment of those essential to the successful design and construction of the project and is responsible for obtaining planning and building regulations consent from the proper authorities.

Fig. 3.2.2 Building project associates

their designed structures have on the public although recently they were given permission to advertise their services in their institute's professional code. They also enter competitions which are organised from time to time by certain prospective clients, such as committees which represent public or other bodies.

Competitions are normally held when a prestige building is envisaged, and the best artistic and functionally sound design which meets the conditions as laid down in the competition organisers' brief is accepted as the winner. Competition winners find that with the publicity given to the winning design, and especially to the structure on completion, they can earn a good reputation.

Architects, as described in Section 2.1, are basically artistically gifted, and initially as students would have usually embarked on an RIBA recognised course of study, provided always that their entry qualifications were suitable. Only professionally qualified and trained architects have the privilege of using the word 'Architect' after their name for business purposes or on correspondence, for others to do so would mean risking prosecution under the Architects' Registration Act 1938.

On structural and civil engineering projects the major part of the design requires complex calculations and that specialism offered only by structural or civil engineers; therefore, only persons professionally qualified in these disciplines are commissioned to design and manage the project. Major structural and civil engineering works, such as: bridges, dams, waterways, power stations, motorways, harbours and dock designs are not normally in the general vein of architects' work.

In employing civil engineering contractors to carry out the construction work the professional structural/civil engineer acts as the overall manager of the contract (see Fig. 3.2.1), preparing drawings, issuing directives, and acting on instructions from the client, in a similar manner to that of an architect on a building project (see Fig. 3.2.2). The resident engineer on-site represents the professional engineer and looks after the interests of the client. He is also known as a clerk of works.

In the design of a structure by an architect specialist professional advice may be sought on such problems as structural stability, soil-bearing capacity, heating, lighting and many other built environment problems. Architects are trained in most aspects relating to the design of buildings, but structural work is so complex that he prudently seeks other professionals' services. For these purposes the following can form the nucleus of the design team, each being employed and paid for directly by the client on the advice of the architect.

The role of the architect is rapidly changing due, in the main, to the newer system of 'management contracting' for multi-million-pound contracts. The client may now employ a firm of management contractors who take control at the feasibility stage, advise on an architect to design the proposed structure and later employ a main contractor to undertake the work.

Structural Engineer

In being commissioned for work by the client this engineer negotiates his fees on the basis of those laid down by the Association of Consulting Engineers. It is an architect's responsibility to convey all structural problems to the engineer who will advise on, or calculate the sizes of, structural components and members. Initially, copies of the 'outline drawings' are issued to the structural engineer to see if the proposed design is feasible, or that the proposed materials are structurally able to withstand the forces on them. Generally the stability of the structure is his responsibility and submissions of calculations to the local authority for approval is necessary at the same time as the architect submits his drawings for building regulations approval by the building controls officer.

Civil Engineer

Advice is sought, and drawings may be needed, where special problems exist regarding reclamation of land, soil stabilisation and ground water. Structural engineers are also trained in these matters, but where the problem is too great, professional civil engineering specialists usually have much greater experience and knowledge on how to deal with them.

Environment and other Services Engineers

In designing modern structures many environmental problems need to be overcome on which trained, experienced specialists are able to give advice. They can also prepare diagrams of their proposals which can then be incorporated within the architect's drawings. Indeed, usually the specialists prepare separate drawings of, say, the services layout to a proposed building, which the architect includes in what are known as, the 'contract drawings'. These drawings are later sent, along with the other 'contract documents', to those companies who wish to tender/bid for the construction work.

Special problems exist when services are to be included in the structure, and services consultants seek advice from the structural engineer, with the architect's approval, on such matters as where best to pass large pipes through floors and walls, and will a particular part of a structure support fixtures for services (water tanks).

Many members of the Chartered Institute of Building Services work in private practice offering their services to architects and designers alike, on such specialisms as follows:

Plumbing: hot and cold water supplies.
Heating: electrical, gas, oil, solid fuel.
Lighting: natural and artificial.
Ventilation: natural and mechanical.
Air conditioning: refrigeration and heating.
Acoustics: insulation, absorption.
Sanitation: above and below ground.
Communications: persons and materials – using lifts, escalators, conveyors.
Refuse disposal: shutes, storage, disposal.
Telecommunications: internal and external systems.

The internal and external comforts of the occupiers of buildings rely ultimately on the sound advice which the aforementioned specialist advisers give to the architect at the design stage.

The Professional Quantity Surveyor

The quantity surveyor is chosen and appointed directly by the client but can be selected on the recommendation of the architect. His fee is usually based on the contract sum.

The quantity surveyor is brought in at the earliest opportunity – normally at the design stage – to advise the client, with the approval of the architect, on the approximate cost of the various schemes put forward to the client. Bearing in mind the cost and other criteria the client will then choose the scheme which is most suitable.

It must be borne in mind that it is unusual to employ a professional quantity surveyor if a Bill of Quantities is not part of the contract documents. Bills are not normally required for small works, schedules being more suitable. Neither are 'bills' prepared by the professional quantity surveyor for contracts offered under a 'package deal', the construction contractor's quantity surveyor being in a better position to do so.

When 'bills' are required and the client requires strict financial control of the project the professional quantity surveyor is appointed, and eventually when the contractor is selected, he keeps in constant touch with the successful tenderer's representative, particularly prior to interim valuations and payments. The surveyor is usually a qualified Associate of the RICS.

Technicians and draughtspeople

Architects may choose to work alongside a partner or an associate. The partner must himself be a qualified architect, while an associate would be expected to have some special skill which contributes to the success of the practice, and therefore some share of the business profits may be offered by the architect as an inducement to retain the services of the associate. An associate, for all intents and purposes, is a type of partner but is not recognised as such by the RIBA, or insurance companies who only give professional insurance cover to those adequately qualified.

In an office of any standing sound supervision is given by a chief technician who naturally controls the technicians/draughtmen and women who execute the detailing work in the office. The technicians and draughtspeople study to HTC or HTD level in architecture, and usually belong to a technicians' institute.

3.3 The contractor and the head office structure

In the industry a firm either operates as the main contractor managing and directing all works on site, or as a nominated or ordinary subcontractor. A nominated subcontractor is the choice of the client or architect, and the ordinary type is that which the main contractor directly employs. The subcontractors offer their services in such specialisms as electrics, plumbing, heating and ventilating, air conditioning, sound and thermal insulation, communication systems, television engineering, refrigeration, internal and external surface finishes, and many more. The organisation structure of some subcontractors is quite sophisticated principally because they also produce the products, which they then fix on-site, in their own factory or assembly plant. They not only have their own construction teams who incorporate their products into the site structures, but also have the 'back-up' services of the design, market research and sales sections, to mention but a few.

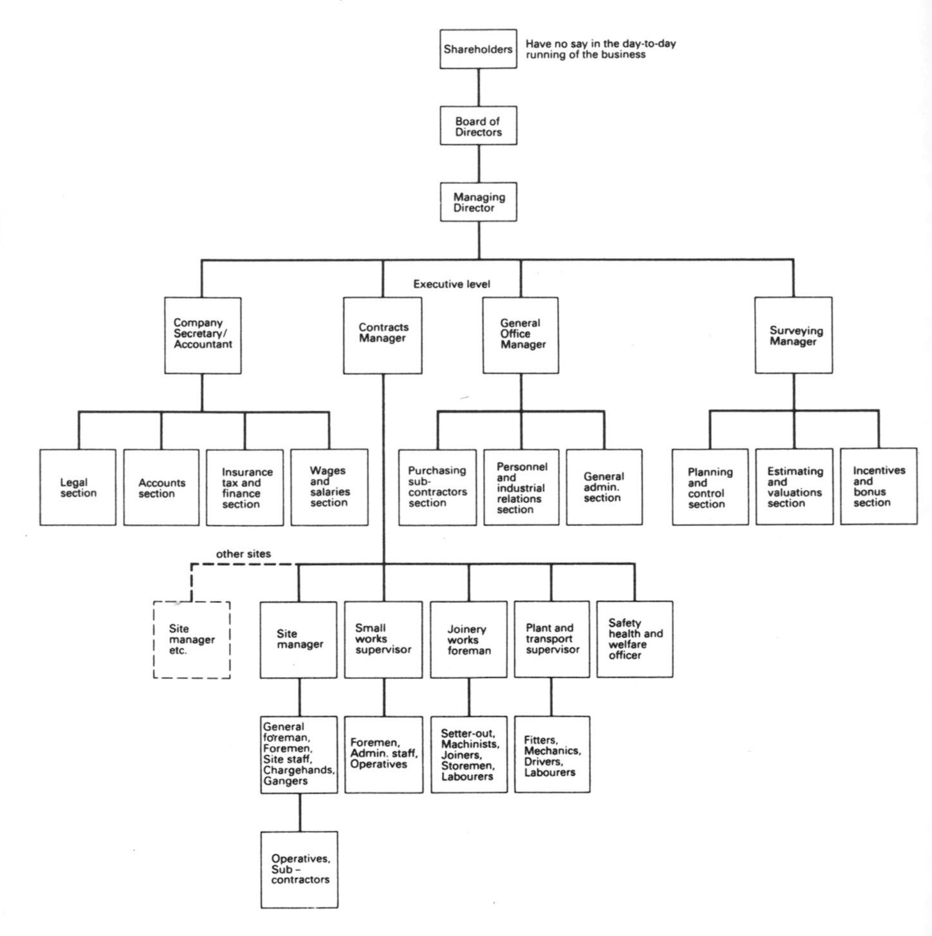

Fig. 3.3.1 Organisation structure of small/medium-sized company.
Note: As the company grows managers' workload can be off-loaded and new posts can be created, i.e.: company secretary/accountant can be divided into two posts. Also, departments such as plant, joinery works, etc. can be made into subsidiary companies.

Building and civil engineering firms or companies can operate on a local, national or international level, and can be divided into small, medium, large and very large concerns. It is usual for the small- to medium-sized firms to operate locally or on a county level, perhaps venturing periodically further afield when economic conditions dictate.

The structure of the many different firms in the United Kingdom varies considerably, but if one considers the structure of a small/medium sized company, and taking cognisance of the work which essentially goes on, the minimum company structure would be similar to that shown in Fig. 3.3.1. It should be noted from the diagram that each executive manager is accountable for more than one section. It would be essential that section leaders (supervisors) would be responsible to these managers, but if, due to the growth of the company, sections become too large, other departments can be created with a separate manager in command of each.

When a company expands considerably and begins operating nationally, adequate planning is necessary to maintain communications between departments, or indeed between the other branches; they should then be based strategically to afford adequate control. The formation of subsidiary companies is sometimes necessary which are then allowed to work almost autonomously to achieve maximum profits. The alternative would be to 'take over' existing established firms for the same purpose. In having subsidiaries and allowing them to work separately from others it gives the managing director more scope to improve the firm's performance. Ultimate control of the subsidiary companies (group of companies), however, would be placed in the parent company, whose main function nowadays is to advise, in a management way, on the running of the subsidiary or satellite companies. Usually one or two directors from the parent company would be voted on to the 'board of directors' of the subsidiary company and would assist in giving suitable guidance to the managing director, and would see that the basic policies of the parent company are implemented.

The 'Group', as the parent and subsidiary companies would be known, could be an international organisation – similar to Britain's very large building and civil engineering concerns – with companies in the manufacturing field, such as: concrete products, industrialised building components and timber products. They also have subsidiary companies in house building, industrial and commercial building (general building), civil engineering, plant and transport, design services, brickworks, gravel pits and numerous others. If one company becomes unprofitable, leading to financial difficulty, it will not normally seriously affect the other member companies of the Group, because it is recognised legally as a separate body.

It should be noted that each subsidiary company normally has its own shareholders, and only by the fact that the parent company is a majority shareholder in each is it able to maintain control over them. Also, a subsidiary company can itself control other subsidiary companies with the main parent company still exercising overall control, as shown in Fig. 3.3.3.

The structure of a medium/large company can be as shown in Fig. 3.3.2. The arrangement of this firm is more suited to the type which operate on a regional basis than on a national one. It shows specialist head office departments each with a manager who is directly answerable to the managing director. The work of the departments is partially shown on the diagram, and with regards to the sections only brief particulars are outlined, as follows, to give an awareness of their existence:

1. Plant and transport section

A plant coordinator or manager would cooperate with the contracts department and would supervise those under his leadership, such as:

(*a*) Plant foreman: supervises and issues instructions to mechanics, fitters, welders, electrical engineers, painters, storemen, labourers, etc.
(*b*) Inspector: inspection reports on incoming plant damage and breakdowns.
(*c*) Drivers.
(*d*) Plant operators.

2. Joinery works section

The works manager directs all operations and personnel in his section. His main concern is to supply cabinets, joinery and carpentry fittings when required. He usually employs 'within the works' the following operatives:

(*a*) Shop foreman.
(*b*) Setter-out.
(*c*) Clerk and storeman.
(*d*) Machinists.
(*e*) Joiners and carpenters.
(*f*) Cabinet makers.
(*g*) Assembly workers.
(*h*) Labourers.

3. Planning section

When a contract is to be tendered for, or has been awarded to the company, the planning section can be called upon by the managers to give advice on the best way to do the work, particularly if the contract is a complex one. Also, programme charts could be prepared to arrive at the contract duration. They also monitor the progress of the contract during construction, as described briefly in Section 3.4.

4. Work study section

Only the larger companies can afford the luxury of a section such as this. Other companies use the services of management consultants if work study or advice is required regarding an organisation problem. The work study officers are usually called in by the contracts manager if improvements are sought to production times and costs. Time studies and method studies are their specialities; in the former, stop watches are used to record elements of work, and method study techniques are sometimes useful to see if a new work method

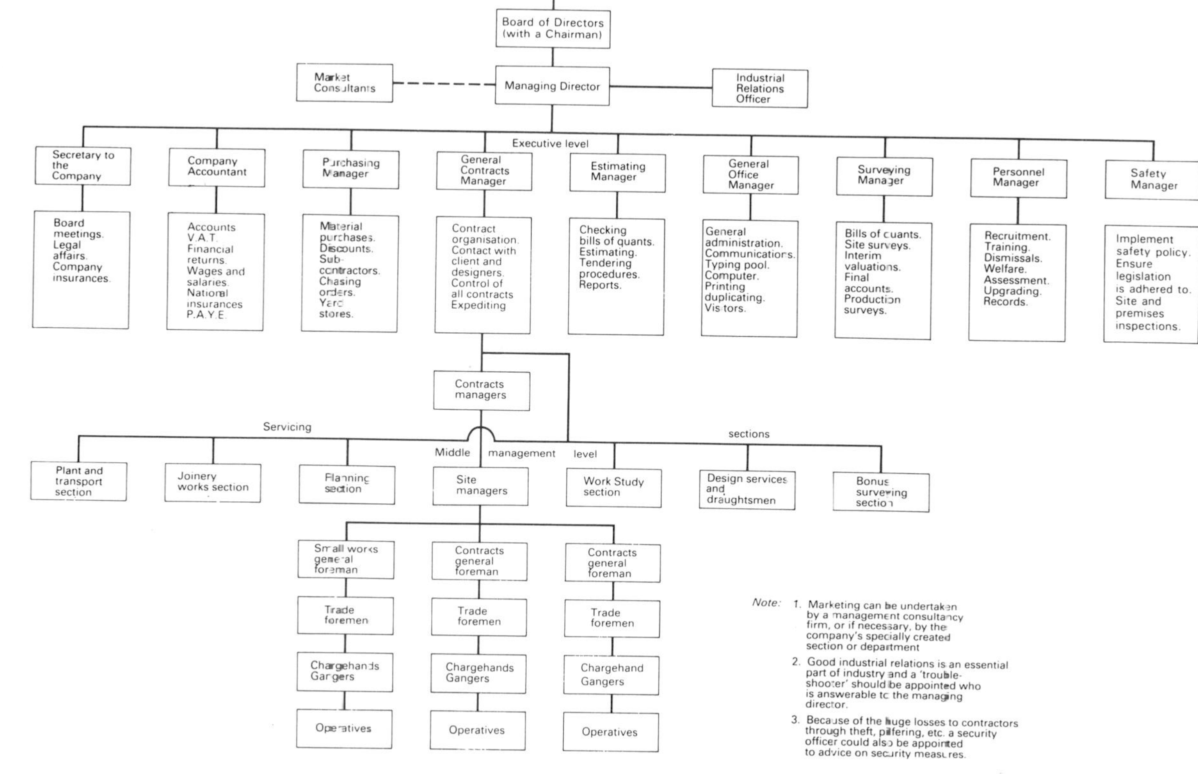

Fig. 3.3.2 Organisation structure of medium/large-sized company

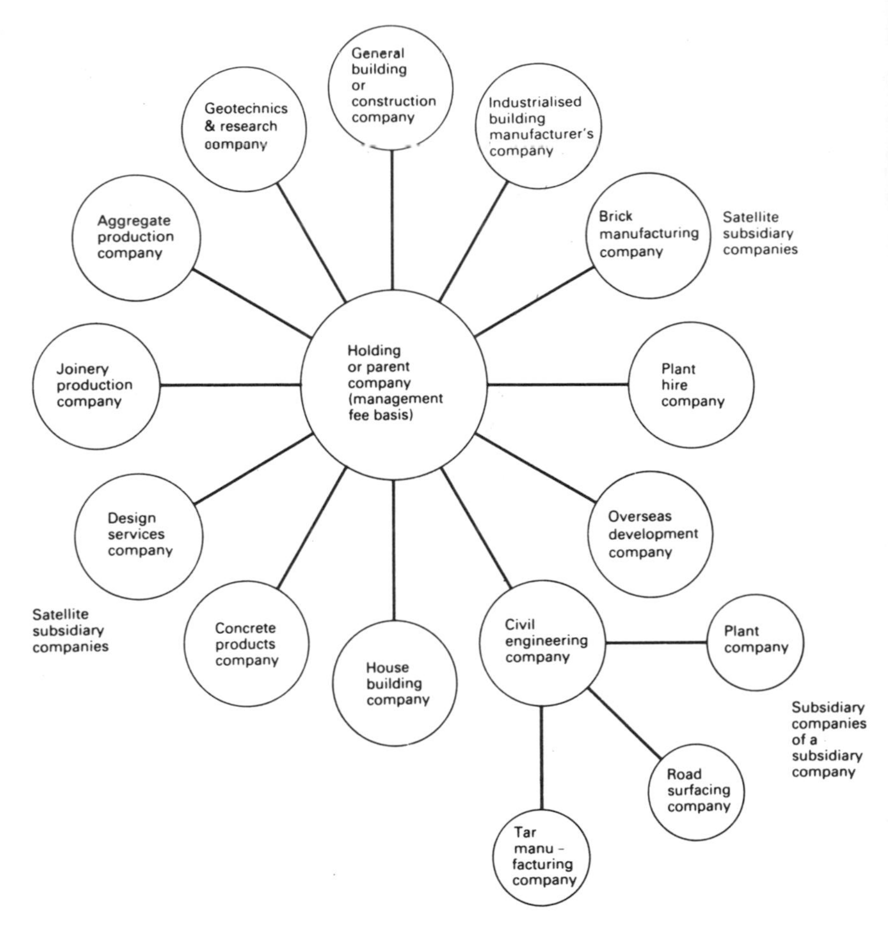

Fig. 3.3.3 Major construction group's organisation structure
Note: This group organisation is unique to a handful of construction concerns operating in Great Britain

can be found to help reduce the production time. Bonus clerks and estimators find the information so obtained from time studies useful to calculate targets for bonus, or for fixing a unit rate for estimating the cost of doing different work. The general manager can use work study officers to carry out a survey of administration methods used in the offices to see if improvements can be made to clerical operations. The joinery works manager may feel that with a little reorganisation of the joinery production line a higher output may result, and the person to assist in a study of this kind, who it is hoped would give recommendations on completion, is the work study officer.

Contracts manager

Other contracts

Other cont acts

Site manager
Safety supervisor

Services sections

Clerks and typist(s)

Storeman and checker

Time-keeper and Wages clerk

Quantity surveyor and planner

Site engineer and bonus clerk

General foreman

Assitant site manager and General foreman

Canteen supervisor and staff

Night security officer

Operatives section

Operatives

Foreman carpenter and joiner

Foreman bricklayer/ mason

Ganger (skilled ops. labourers)

Crane, hoists and other plant operators

Scaffolders

Sub contractors

Steel erectors

Window fixing specialists

Plumbing contractors

Electrical contractors

Chargehand C and J carcassing and 1st fixing

Chargehand C and J 2nd fixing

Chargehand C and J formworking

Chargehand bricklayer

Chargehand bricklayer

Chargehand bricklayer

Steel fixers

Concrete gang: mixer driver barrowmen spreaders and layers

Excavator and Dumper drivers, banksmen and gen. labourers

Plastering and ceiling finishes contractor

Heating and ventilating contractor

Gas installations

Painting and glazing contractor

Tradesmen apprentices labourers

Tradesmen apprentices labourers

Tradesmen apprentices labourers

Tradesmen apprentices labourers

Tradesmen apprentices labourers

Tradesmen apprentices labourers

Sewer and drain layers

Road and pavement layers

Floor and wall tile contractor

Roof finishing contractor

Landscaping and fencing contractor

Note: 1. This figure takes into consideration the anticipated work-force expected on the site over a two-year period.

2. Subcontractors are to use their own foreman for their trade supervision.

3. This site structure indicates the site personnel delegated to each supervisor.

4. Building structure is partly of steel frame the remainder being of reinforced concrete frame.

5. This figure does not consider the use of labour-only subcontractors.

Fig. 3.4.1 Contractor's site team (factory and administration block under construction)

 5. *Design services section*

If a company operates a 'speculative' construction section (develops land owned by it) designs are necessary from which the construction teams are to work. Instead of using the services of outside architects and engineers the company can employ its own team permanently. They can also be used if the company negotiates a 'package deal' contract with a client. This is to say, that the company will design and build the structure for an all-in fee.

6. *Bonus surveying section*

It is the responsibility of this section to fix, usually by work-study methods (time study), or negotiate with the operatives or their union representatives a satisfactory 'rate' for doing various work. The rate should be such that operatives, if working at an average speed, can earn bonus equivalent to one-third of their normal pay. The advantages of this practice are: The operatives tend to waste less time by arriving promptly for work, rest periods are minimised and work is done more efficiently. Naturally the work is completed more quickly and the contractor can then use his resources earlier on other contracts; as a result, the client is able to take charge of his building on time. With these very evident advantages the cost of a bonus scheme is justified.

Note. According to the 'Working Rules' a bonus scheme should be operated by all construction firms. Many firms satisfy these requirements by paying a standard bonus of between £10 and £20 per week whether the operative earns it or not.

7. *Safety Manager/Officer*

Sites, offices and other work areas are visited by this officer who advises on the best ways to satisfy the Health and Safety at work, etc. Act 1974. Reports are prepared after each inspection to keep checks on whether recommendations have been implemented or ignored. He also liaises with the site manager and union representatives if there are any problems which are highlighted regarding safety, health or welfare (see also Chapter 7).

3.4 Site structure and organisation

The organisation structure for any site varies considerably depending on the size of the construction firm and the type of contract work being undertaken. While there is a pattern operated by some building or civil engineering contractors no site structure is exactly the same. The usual pattern, however, is for the contract manager to take control of the contracts under his direction and that his site managers – sometimes referred to as site agents, site supervisors, contract controllers, or project managers, depending on the firm for whom they are employed – are responsible for the day-to-day running of the contracts in their charge (see Fig. 3.4.1).

For sound organisation, particularly on a large contract, an assistant to the site manager (general foreman) is essential and he will issue instructions to the site operatives through their foremen and gangers, on behalf of, and as directed by, the site managers. Also, control of the works would be the assistant's responsibility in the absence or incapacitation of the site manager. The assistant site manager in numerous companies has usually progressed from tradesman, chargehand and foreman to the position of general foreman. This qualifies him to take on the responsibility of assistant site manager, or where this is not possible on large contracts, he would at least be delegated to direct the operations of more than one trade. The alternative route to the post of an assistant or site manager would be by being company trained after qualifying through a university or higher education establishment.

Trade foremen take instructions from the general foremen who are their immediate superiors, and in turn allocate work and direct the operations of the respective gangs of tradesmen, with the assistance of their chargehands. The bricklaying foreman, however, does not normally supervise the work of carpenters and joiners, neither would the joinery craft foreman ever supervise bricklayers.

It is usual on a medium-large contract to have the resident services of the quantity surveyor, planner, site engineer and others to assist in the administrative, cost, quality and progress control. It also helps to ease problems regarding bonus payments if the bonus clerk is readily available to discuss matters with aggrieved operatives.

The work of some of the personnel on construction sites is briefly described as follows:

1. Site Manager

The Site Manager has the complete responsibility for all operations and personnel on the site. Also, visitors must have his approval before venturing on to the site whether it is in a professional capacity or simply as inquisitive outsiders interested in the project. Therefore, he is:

(*a*) subordinate to the contracts manager;
(*b*) superior to other personnel on the site;
(*c*) to carry out the policies of the company;
(*d*) to control the site work programme using site meetings and the contract programme;
(*e*) responsible for directing the administration and communications such as: site diary, telephone, weekly returns;
(*f*) responsible for industrial relations – negotiations with union representatives when necessary;
(*g*) to liaise between the client's representatives and the contract manager or head office;
(*h*) site safety supervision, inspection records, notifications.

These are but a few of his responsibilities.

2. Site Engineer

The Site Engineer takes instructions from the site manager and is responsible for:

(*a*) setting out the framework of the structure accurately from the plan drawings for all excavation work, concreting, brickwork, steelwork, drains and road;
(*b*) setting out levels as and when required;
(*c*) assisting in the quality control of concrete and other work, and the keeping of records;
(*d*) constantly checking levels and verticality of the structure as work proceeds.

3. Quantity Surveyor

The Quantity Surveyor is subordinate to the site manager, but more usually is directed by the chief surveyor or surveying manager at head office, and:

(*a*) measures the work done on-site in order that an acceptable figure can be agreed with the client's professional quantity surveyor;
(*b*) records variations to the work shown in the 'Bills' by the architect, so that claims can be made for payment;
(*c*) prepares final accounts, taking into consideration other costs such as day works and fluctuations;
(*d*) records and agrees the work done monthly with the various subcontractors;
(*e*) is responsible for cost control information being sent to head office.

4. Planner/Programmer

(*a*) assists the site manager to prepare weekly and/or monthly programmes of work for the various trades and subcontractors, using the master programme as a guide;
(*b*) up-dates the master programme (bar chart or critical path chart) if and when required. The programme needs up-dating when the contract is behind schedule and requires condensing to meet the contract completion date. The methods normally adopted to reduce the contract time are: allow overtime working, use more plant and labour, or introduce quicker and more efficient techniques;
(*c*) records the various stages of work completed so that progress can be monitored by the contracts manager.

5. Time-Keeper and Wages Clerk

(*a*) Records arrival and departure of the site personnel;
(*b*) prepares wages sheets, using:
    i. Tax deduction cards;
    ii. Time sheets or records;
    iii. Bonus slips (from bonus clerk);
    iv. PAYE tables A, B, C, or D, etc.;
    v. Insurance tables.
(*c*) arranges for the collection of cash from the bank;
(*d*) inserts cash and pay slips into pay packets;
(*e*) distributes pay at appropriate time taking care to observe the company security code against robbery.

6. Storeman and Checker

(*a*) ensures adequate materials and expendable items are in stock;
(*b*) calls forward goods when stocks are low and checks deliveries and signs delivery notes.
(*c*) records distribution of materials;
(*d*) issues small plant and records returns;
(*e*) is sometimes responsible for providing storage space for delivered goods, etc.

7. Clerk and Typist

(*a*) receives and opens non-personal incoming mail;
(*b*) distributes to appropriate offices or sections;
(*c*) collects, francs and posts outgoing mail;
(*d*) prepares letters and files copies;
(*e*) other general administration such as: recording, indexing and filing, duplicating and dealing with incoming telephone calls.

8. Canteen Supervisor

The responsibility for supplying meals on-site is sometimes subcontracted to an individual or firm. Canteen facilities (a room, equipment, fuel) are normally made available in these instances by the main contractor, and the buying-in, preparation and the selling of food and drinks is the responsibility of the canteen manager/subcontractor.

Usually only snacks are provided to minimise costs.

9. Ganger

The Ganger acts in a similar capacity to that of a foreman, but deals primarily with the semi-skilled (concretors, plant operators) and unskilled labour (general labourers). While working for some contractors he may be called upon to direct the work of the skilled workers (steel fixers, drain layers, paviors) but not the craftsmen (tradesmen).

While the remaining operatives' and other personnels' responsibilities and type of work is self evident, as shown in Fig. 3.4.1, it is necessary to state briefly the work of the carpenters and joiners. Their site work is divided into four sections, as follows:

1. *Carcassing work:* • The fixing of roofs, floors and other structural parts necessitating, normally, the use of timber, but other materials could be incorporated such as: steel, aluminium, glass fibre.

2. *First fixing:* The work which is carried out after the carcassing parts are completed and before the plastering operations, if there are any, are commenced. These are: door frames, fixings and backings to panels.

3. *Second fixing:* Refers to the joinery finishes, usually after plastering, which are fixed prior to painting, i.e. doors, skirtings, architraves, cupboards, ironmongery and furniture.

4. *Formworking:* The preparation of temporary boxes, decking or upstands, to support fresh batches of concrete until setting takes place and an adequate strength is reached. Formwork may therefore be used for beams, floors, columns, stairs, or any structure facilitating the use of concrete. Other names given to formwork are shuttering or falsework.

3.5 Visitors to site

During construction work on-site the site agent or manager could be constantly interrupted from his supervisory role regarding control of all operations and workpeople on site. He has other responsibilities which at first are not apparent to others in a less responsible position, or the layman. He is not only in charge of all activities but is an ambassador of the company/firm for whom he works. He usually needs to be tactful when necessary, and articulate to the point that all those with whom he becomes associated will have respect for him and his position on-site. Most site managers have a flair for getting on with people from whatever social background they originate, and because of his personality is able to cope with the regular stream of visitors who essentially need to enter on to the site, in addition to those who are inquisitive and find the project of great interest. The main visitors, other than the contractors' employees, are described as follows:

1. The Client

The Client usually visits the site to see what progress is being made, and should not by-pass the architect by giving instructions to the contractor's representative. If instructions are necessary, then the architect must be notified immediately and he will arrange to advise the contractor of the client's requirements.

If the client insists on a personal approach, requesting the site manager to make a variation, the manager should confirm the request to the architect by letter.

2. The Architect

(*a*) checks progress of work on-site;
(*b*) answers queries where necessary;
(*c*) agrees variations to the contract documents, namely, alterations to the design and specifications;
(*d*) attends site meetings periodically to help inject impetus to the project, and discuss delays and other problems which may be critical;
(*e*) checks quality of work;
(*f*) inspects samples of materials and fittings delivered to the site;
(*g*) conducts the client round the site when necessary.

3. The Engineer

He answers queries on-site if requested by the architect, and checks on structural points when problems arise.

4. Other Consultants

They may visit the site with the architect to discuss pertinent points regarding the project.

5. Professional Quantity Surveyor

He visits the site to deal with the following:

(*a*) to prepare interim valuations of contractors' and subcontractors' work so that a certificate for the value can eventually be submitted by the contractors to the client for payment;
(*b*) checks claims for extra payments which are made for additional work done by the contractor;
(*c*) continuously checks on project costs for the architect;
(*d*) at 'practical completion' (when building is ready for occupation) valuation measurements are taken and a 'certificate' is prepared;
(*e*) when faults have been made good, usually six months after employer/client has occupied the structure a 'Certificate of Completion of Making Good Defects' is issued to the contractor;
(*f*) 'final accounts' are prepared at final completion (after 'defects liability period' and 'Certificate of Completion of Making Good Defects') to arrive at the sum still owing to the contractor.

6. Building Controls Officer

The site is visited by this officer at various stages of the project. Inspection of the structure is made to see that it conforms to the Building Regulations 1976, and amendments, and the drawings which were submitted to his/her office at the Building Regulation Approval stage by the architect. (There will be new Building Regulations 1985).

7. Clerk of Works

The Clerk of Works is employed by the client and therefore represents his interests on-site. His total responsibility is with regards to the specifications prepared by the architect to which the contractor should adhere. The inspection for quality of the project lies with him. On large projects the clerk of works would be resident on the site, and accommodation would be made available in the shape of a hut or room and furniture for his administrative responsibilities. On small works he would visit the site at regular intervals to inspect structural and other critical work, comparing it with the 'specifications' and the rest of the contract documents. Brickwork, concrete work and work executed by all trades on-site are scrutinised by him which ensures that the contractor supervises operations carefully at all stages.

Records are kept by the clerk of works of all events, such as stoppages, delays, weather and anything which when reported will assist the architect when 'extensions of time' and other claims are made by the contractors.

8. Other Local Authority Technical Officers

Highway engineers, public health inspectors and town and country planning officers may make spot checks to parts of the work relevant to each representative's delegated powers on such points as road access, nuisances, infringements and other types of complaints made by neighbours-to-the-site and the public generally.

9. Subcontractors' Managers

Their visits usually coincide with the site meetings which are either called by the architect or the main contractor. They may, however, be called on to site when problems develop regarding progress or labour, which is outside the main contractor's powers.

10. Public Utility Service Organisations (gas, water, electricity, etc.)

They are usually in the same category as subcontractors but their representatives may check on such problems as damage caused to their existing underground pipes or overhead cables by the contractor's excavators or cranes.

11. Factory Inspectorate

The inspectorate examines the way that safety, health and welfare facilities are maintained on-site. Inspectors, who are employed by the Government, have right of access to any site, office, factory, etc. They visit construction sites to check the various inspection records and accident books. The legislation relevant to site safety, etc. is dealt with in Chapter 7.1.

12. Suppliers

Nominated or ordinary suppliers of materials and other products visit the site to discharge their loads. Sound organisation is essential and areas should be prepared for storage, and labour should be made available usually for off-loading to eliminate delays.

13. Police

They may wish to inspect the site during the day or night particularly if a request is made by the contractor as a security measure. Also, they are evident when problems arise such as obstruction of the highway adjacent to the site by contractors' vehicles, skips and materials. Unloading periods on busy highways sometimes requires the services of a policeman and an approach should be made, when necessary, to the appropriate authority.

14. Union Officials

Full-time union officials have the right, under the Working Rule Agreement, to reasonable access to the site to inspect union cards, collect union subscriptions and to discuss with the part-time union representatives any grievances which may have arisen between different trades, between different unions or themselves and management.

15. Fire Officer

This officer may wish to inspect the site for fire hazards, particularly regarding the building under construction, and to see that the fire recommendations made during 'design stage' are being implemented successfully into the structure.

16. Special Visitors

Under this category would be placed the following: historical buildings and amenities groups, women's institutes, students of architecture or construction, professional institutes, councillors, foreign visitors and many more, With these types of visitors the site manager would have to check that proper insurance cover is maintained by his company. Subject to this being satisfactory then correct arrangements must be made to enable the visits to progress smoothly. Adequate notice must be given to the site manager by intending visitors and then the following arrangements would be adequate.

Prior to visit

(*a*) Acknowledge the letter received from intending visitors, if it has not already been done by head office, stating expected time of arrival and directions if necessary. Also, name the person to whom they should report.
(*b*) Check site tidiness and safety.
(*c*) Erect direction signs along suitable safe route, if necessary.
(*d*) Arrange parking facilities.
(*e*) Nominate a guide, if site manager is not conducting the site tour himself.
(*f*) Arrange for tea and biscuits and a clearing-away service.
(*g*) Supply protective clothing, if necessary, and safey helmets essential.
(*h*) Spare room to be made available to introduce the guides and to outline the various facets of the project. The route which will be taken is outlined, and emphasis should be placed on safety.
(*i*) A model of the project should be available.

On arrival of the visitors great care should be taken regarding their welfare and safety. At all times it should be stressed that they keep to the planned route and should not stray from the main party.

Special Note: As the site manager is responsible for everyone and everything on site, all visitors, no matter who they are, should report to the main office before progressing further.

Chapter 4

Design principles and procedures

4.1 Sources of information

Once the initial approach has been made to the architect by the client to undertake the design of a structure the architect will require a range of information to which he can refer regularly. This is known as his 'back-up' service and will be used to contribute to that information already gained by him through his educational training, and above all, through practical experience during his period in professional practice. These resouces are essential so that correct sizes, materials, components and the whole range of ergonomic requirements are met when designing. Knowledge of the existence of such documents as the Town and Country Planning Acts, Public Health Acts, Housing Acts, Highways Acts, Local Government Acts, Clean Air Acts and Regulations such as the Building Regulations and Amendments are also required. So much of the Government's legislation affects building and development generally that the architect needs to update his knowledge regularly, particularly when a new law affecting development is passed by Parliament.

The greatest need of an architect's practice is to have a suitable library system. The way in which the system is designed and organised will generally reflect the efficiency of the organisation. The most usual method of indexing information, in addition to the alphabetical or numerical systems of filing, is by the system known as the CI/SfB System. It is particularly suitable for filing information on new products of materials or components, but because of its flexibility can, in addition, be used to file almost any information from details on a type of building, to the financial implications of a building. Drawings, bills of quantities and other contract details of past projects can be filed within such a system. However, correspondence and other communication information may be filed for convenience, either by job number, or alphabetically under the heading of 'communications'.

The SfB System was introduced on licence into this country in 1961 by the RIBA. The system was first created in Sweden by what was then known as 'Samarbetskommittën for Byggnadsfrågor' and named after the Swedish Technical Secretariat of the Coordinating Committee for Building. So the name Samarbetskommittën for Byggnadsfrågor, hence SfB, was used for the system. The system is now universal, and the United Kingdom has gone one stage further by introducing the CI/SfB system (CI = Construction Index) in 1968, which was then revised in 1976.

There are firms in existence which were created to deal exclusively with the dissemination of building and civil engineering information. Their services are sought by organisations and businesses which require to maintain up-to-date information on new materials and products generally. Architects, contractors, local authorities and further educational establishments normally have a reference library to which they can refer when information is needed. Firms such as Barbour Index Limited, Building Products Limited and the Royal Institute of British Architects Office Library Service collect literature from manufacturers' publications and various other sources to incorporate into their clients' reference libraries. The service is financed by the organisations who require their libraries up-dating periodically.

An ordinary numerical or alphabetical filing system is normally unable to cope adequately with all the information relevant to construction work, whereas the CI/SfB system has extendable sections to deal with the mass of information necessary to feed firms' requests for back-up data.

An index book, known as the *CI/SfB Construction Industry Manual* (published by RIBA Publications Ltd.), is retained by those setting up or already operating this system of filing. It is an essential part of the system, and assists the operators to index and file information sensibly. When retrieval of data is necessary the user normally refers to the 'index book' which will give the location of where the material is stored. Folders, files, boxes or pigeon holes are used in order to store information neatly, and should be marked to assist users during their search through the system.

Manufacturers of construction materials and components and writers of articles who anticipate that their information will be filed obtain the appropriate letters or numbers allocated by the CI/SfB system from the 'index book' and incorporate them on to a standard box drawn or affixed at the top right-hand corner of the literature or document (see Fig. 4.1.1).

It is necessary, therefore, to know what sections the system offers. In the SfB system it is set out in three main tables, i.e.:

Table 1 – Functional Elements.

Table 2 – Construction.
Table 3 – Materials.

In the CI/SfB revised system there are two further tables and modifications to the three already shown, which are: Table 0 and Table 4.

Table 0 – Physical Environment.
Table 1 – Elements.
Table 2/3 – Construction Form and Materials.
Table 4 – Activities and Requirements.

Table 0

The Physical Environment refers to the construction end product, i.e. type of building. It is sub-divided into ten sections headed as follows:

0 Planning areas.
1 Utilities, civil engineering facilities.
2 Industrial facilities.
3 Administrative, commercial, protective service facilities.
4 Health and welfare facilities.
5 Recreation facilities.
6 Religious facilities.
7 Educational, scientific, information facilities.
8 Residential facilities.
9 Common facilities, other facilities.

Each section of this table is further sub-divided, for example:

7 = Educational, scientific, information facilities.
71 School facilities.
72 Universities, colleges, other education facilities.
73 Scientific facilities, etc.

Table 1

Elements refers to the parts of a structure, such as walls, floors, ceilings, services. This section is also sub-divided into ten sections but the numbers are in parentheses.

(1-) Ground substructure – substructure.
(2-) Structure, primary elements, carcass.
(3-) Secondary elements, completion of structure.
(4-) Finishes to structure.
(5-) Services, mainly piped, ducted.
(6-) Services, mainly electrical.
(7-) Fittings.
(8-) Loose furniture equipment.
(9-) External elements, other elements.

These sections are sub-divided as before, for example:

(2-) = Structure, primary elements, carcass.

- (21) walls, external walls.
- (22) internal walls, partitions.
- (23) floors, galleries, etc.

Table 2

Construction Form refers to bricks, blocks, pipes and tiles and are sub-divided from A to Z.

A	Construction, forms.
B	Demolitions and shoring work.
C	Excavations and loose fill work.
D	Vacant.
E	Cast in-situ work.
F	Blockwork, brickwork.
G	Large block, panel work.
H, I, J	Sections — sections, pipe work, Wire work, mesh work.
K—V	Sheets — quilts, flexible sheet work (proofing), malleable sheets, rigid sheet overlap work, thick coating work, rigid sheet work, rigid tile work, flexible sheets, film coating and impregnations work, planting work, work with components, formless work and joints.

This table is not sub-divided but may be used in conjunction with Table 3 which deals with the actual material of the construction form.

Table 3

Materials are referred to in this table and each are allocated letters (in lower case).

a	Materials.
b, c, d	Vacant.
e	Natural stone.
f	Precast with binders.
g	Clay (dried, fired).
h	Metal.
i	Wood.
j	Vegetable and animal fibres.
k	Vacant.
l	Not used, causes confusion with i.
m	Inorganic fibres.
n	Rubbers, plastics, etc.
o	Glass.
p	Aggregate, loose fills.
q	Lime and cement binders, mortars, concrete.
r	Clay, gypsum, magnesia and plastics binders, mortars.
s	Bituminous materials.

t Fixing and jointing materials.
u Protective and process/property modifying materials.
v Paints.
w Ancillary materials.
z Substances.

As in Tables 0 and 1 these sections are sub-divided, for example:

i = Wood.
 i1 Timber.
 i2 Softwood.
 i3 Hardwood.
 i4 Wood laminates.

Table 4

Activities and Requirements applies to anything which results from the building process.

(A) Administration and management activities, aids.
(B) Construction plant, tools.
(C) Vacant.
(D) Construction operations.
(E) Composition.
(F) Shape, size.
(G) Appearances.
(H) Context.
(I) Not used.
(J) Mechanics.
(K) Fire, explosion.
(L) Matter.
(M) Heat, cold.
(N) Light, dark.
(P) Sound, quiet.
(Q) Electricity, magnetism, radiation.
(R) Energy, side effects, compatability, durability.
(S) Vacant.
(T) Application.
(U) Users, resources.
(V) Working factors.
(W) Operation, maintenance factors.
(X) Change, movement, stability factors.
(Y) Economic, commercial factors.
(Z) Peripheral subjects: form of presentation, time, space.

As before, sub-divisions occur which, for example, are shown for (P).

(P) = Sound, Quiet.
 (P1) Noise types, sources.

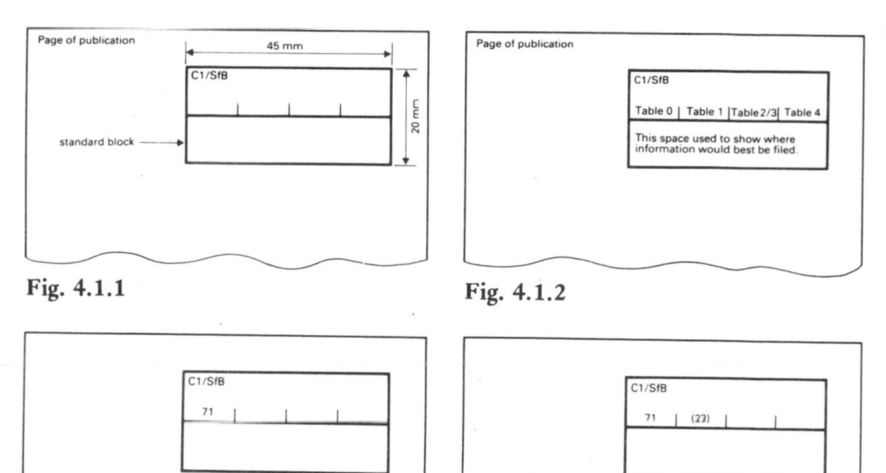

Fig. 4.1.1

Fig. 4.1.2

Fig. 4.1.3

Fig. 4.1.4

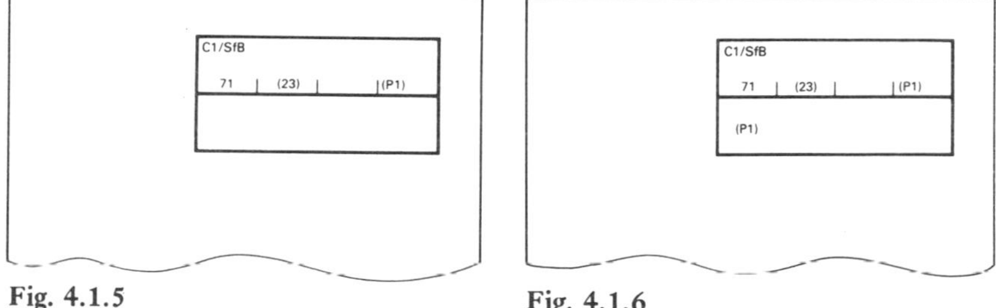

Fig. 4.1.5

Fig. 4.1.6

(P2) Sound insulation.
(P3) Sound transmission.
(P4) Sound reflection, etc.

Technical writers are advised, after they have prepared a written article, to establish the reference number or letter from the *CI/SfB Index Book*. The number/letter should then be incorporated within a standard block at the top right-hand corner of the article or document to enable the librarian/clerk to file the information easily into their existing system (see Fig. 4.1.1).

The block can be divided into the upper and lower parts. The former is used to classify information in the order of Tables 0, 1, 2, 3 and 4. Publications which are to be classified under Tables 2 and 3 are allocated the third space along the top. The lower half can be put to use showing the most convenient section in which to file information where there are insufficient copies to cross reference fully (see Fig. 4.1.2).

In order that one can appreciate how the CI/SfB system works, an example is shown:

Consider a piece of literature which has been received and needs to be classified and filed in the appropriate section of the reference library. The literature refers to 'Sound insulation of floors to schools'.

A question is asked relating to each table in the following order:

Question 1

Does the subject in the piece of literature relate to Table 0 – type of building?

In this example the answer is yes, because the type is a school. In Table 0 schools would be found under section number 7 which deals with Educational, etc. facilities or buildings. In this sub-division 7.1 deals with school facilities. Hence, in the first box which has been reserved for Table 0 details, the figures 71 would be inserted (see Fig. 4.1.3).

Question 2

Does the subject relate to Table 1, i.e. an element of a building, such as floors and walls?

In this example it does because it mentions floors.

In Table 1 floors are allocated the number (23) (see Fig. 4.1.4).

Question 3

Does the subject relate to Table 2 or 3, i.e. construction form (bricks, tiles) or materials?

As no reference is made to either of these the third block is disregarded.

Question 4

Does the subject relate to Table 4, i.e. activities and requirement?

In this example sound insulation is a requirement and can be found under (P1) (see Fig. 4.1.5).

Retrieval

When attempting to retrieve information from the box files the alphabetical section of the index book should be used. Take, for instance, the need to find information on foundations. In the alphabetical section it is necessary to search for the word 'foundation' and beside the word will be given the indexing code; this shows in which box number/letter to look.

As the information is classified under the Tables 0, 1 and 4, photocopies can be made and each would be filed in the three sections of the filing system, if necessary. On the other hand the subject 'sound insulation to floors' is important as there is very little difference, if any, in the construction of sound insulation of floors to schools, flats or office buildings. The information need only then be filed under 'sound insulation' unless the construction details of the floor are worth investigation. The CI/SfB classification block will therefore have printed on the lower part the important section in which the information should be filed (see Fig. 4.1.6).

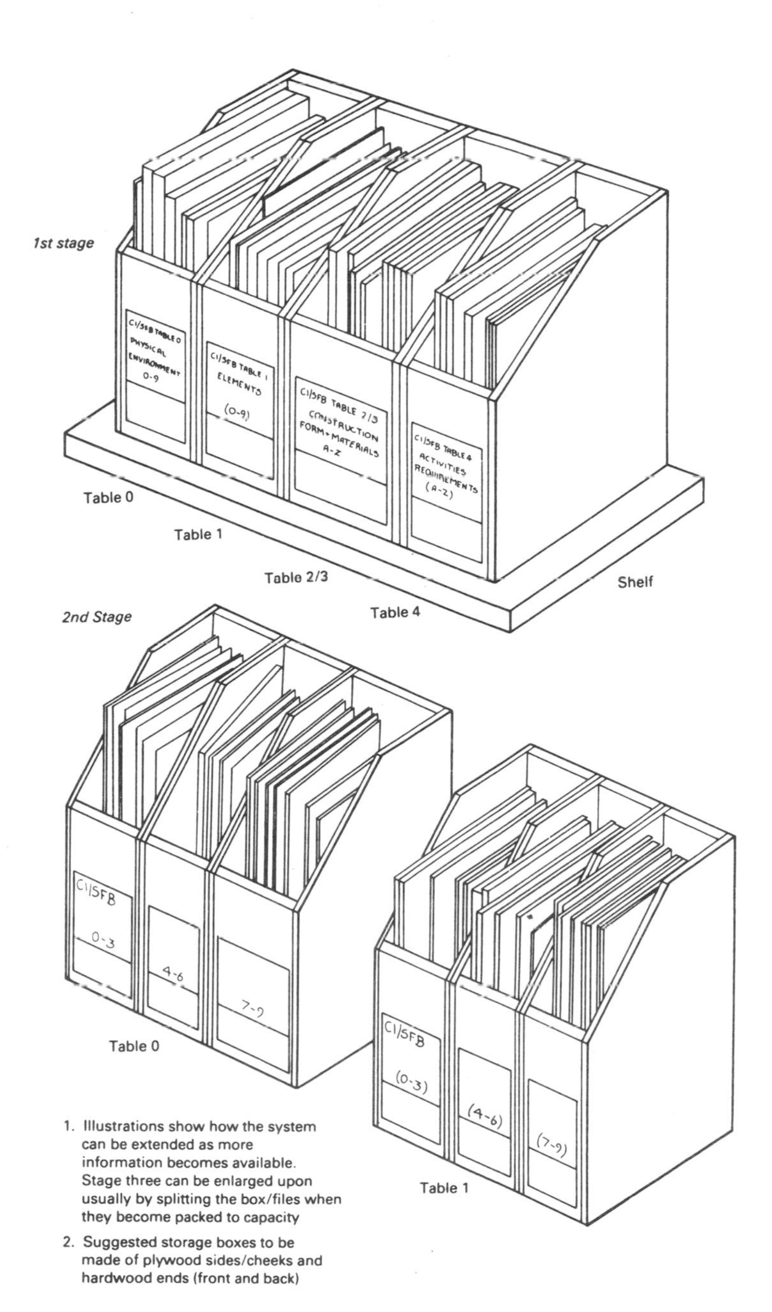

Fig. 4.1.7 CI/SfB filing system layout

Setting up the system

Boxes, files, folders or pigeon holes can be used in which to file information.

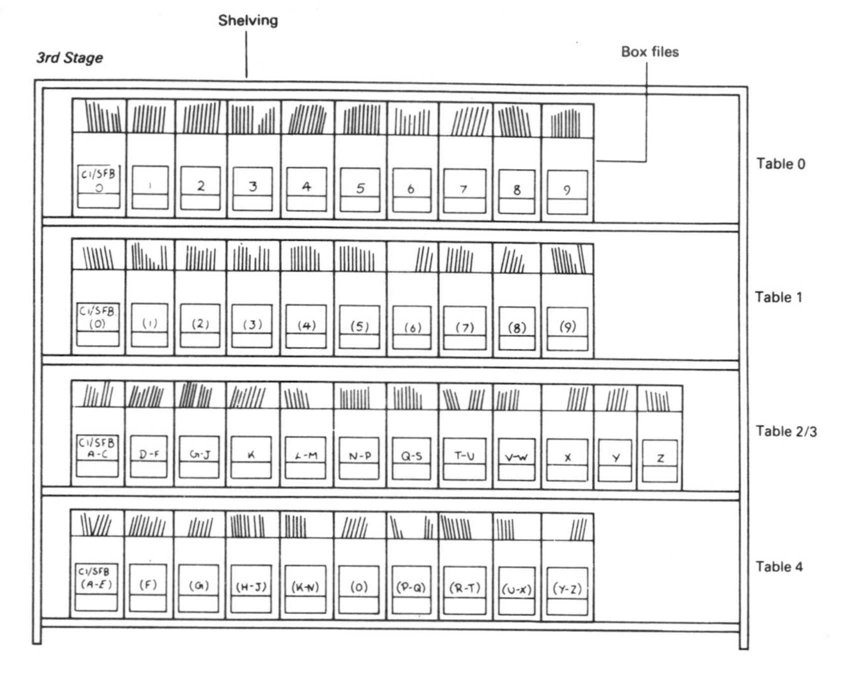
Shelving
3rd Stage
Box files
Table 0
Table 1
Table 2/3
Table 4
CI/SFB 0
1
2
3
4
5
6
7
8
9
CI/SFB (0)
(1)
(2)
(3)
(4)
(5)
(6)
(7)
(8)
(9)
CI/SFB A-C
D-F
G-J
K
L-M
N-P
Q-S
T-U
V-W
X
Y
Z
CI/SFB (A-E)
(F)
(G)
(H-J)
(K-N)
(O)
(P-Q)
(R-T)
(U-X)
(Y-Z)

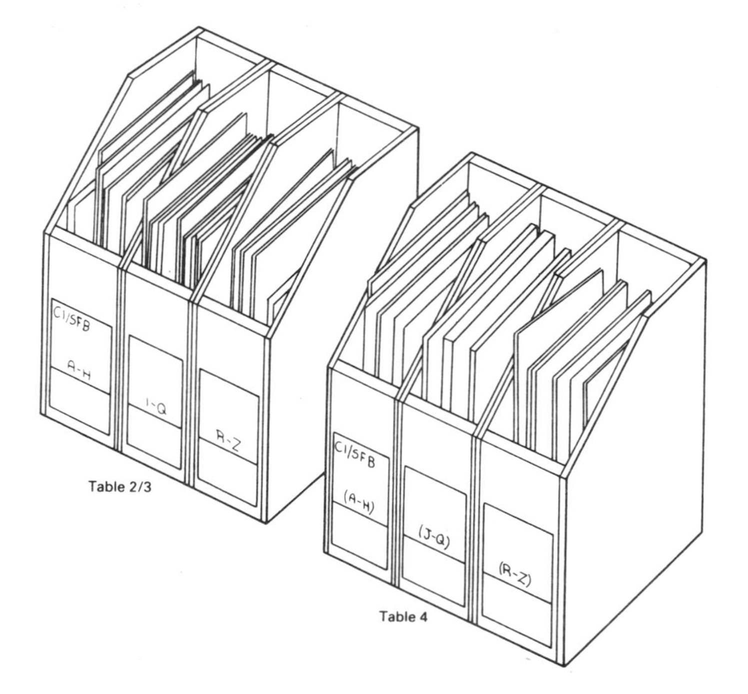
CI/SFB A-H
I-Q
R-Z
Table 2/3
CI/SFB (A-H)
(J-Q)
(R-Z)
Table 4

Boxes are most advantageous because if they are reasonably designed the information so contained within can be seen at a glance, which always assists a user who wishes to retrieve information quickly. In addition to the storage method, the *CI/SfB Index Manual* is indispensable.

One box is used for each section of the system when it is first established. The numbers can then be extended as required (see Fig. 4.1.7).

Sources

Architects, while establishing this important and essential system, would next require to collect other information in the form of handbooks, journals and guides from which to refer when special problems are faced. The types of information are as follows:

1. *Names and addresses*

(*a*) Consultants – quantity surveyors, structural and civil engineers, electrical and mechanical engineers, and other services engineers.
(*b*) Main contractors.
(*c*) Subcontractors and suppliers.
(*d*) Research and other organisations, including trade associations.
(*e*) Local authority departments of the districts in which the architect operates.

This information can usually be found by consulting the membership handbooks of the organisations concerned, such as Building Employers Confederation, Federation of Master Builders, Federation of Civil Engineering Contractors, Royal Institute of Chartered Surveyors, Institute of Civil Engineers, Institute of Structural Engineers, Chartered Institute of Building Services, Institute of Clerk of Works.

2. *Professional Institutes' Manuals/Handbooks*

(*a*) *RIBA Architectural Practice and Management*, by RIBA Publications Ltd.
(*b*) *RIBA Architectural Job Book.*
(*c*) *Architects' Journal (AJ) Metric Handbook.*
(*d*) *Chartered Institute of Building Services' Guides on Services Design* – Book A, B and C.
(*e*) *Specifications* (Architectural Press).
(*f*) *CI/SfB Project Manual* (RIBA). Deals with organisation and building projects information.
(*g*) *National Building Specifications.*

3. *Legal documents*

The *Architects' Journal (AJ) Legal Handbook* is necessary for all architects in professional private practice, and it would be prudent to have available the relevant Acts of Parliament and Regulations, examples of which are as follows:

(*a*) Town and Country Planning Acts 1959–1972.
(*b*) Public Health Acts 1936 and 1961.
(*c*) Clean Air Acts 1956 and 1968.

(*d*) Highways Acts 1980.
(*e*) Housing Acts 1957–1980.
(*f*) Building Regulations 1985 and amendments.
(*g*) Local Land Charges Act 1975.

4. *Manufacturers' Handbooks*

The more common handbooks being for door and window components, plumbing, concrete materials and components, industrialised units or systems, and floor finishes.

5. *Drawings and details of past projects*

Past project information can be useful for references.

(*a*) Working drawings.
(*b*) Bills of quantities.
(*c*) Specifications.
(*d*) Photographs.
(*e*) Reports.
(*f*) Costs analysis sheets.
(*g*) Correspondence.
(*h*) Schedules.

6. *Government publications*

Normally these can be obtained through Her Majesty's Stationery Office (HMSO).

(*a*) Department of the Environment – Advisory Leaflets.
(*b*) Property Service Agency – Schedule of Rates.
(*c*) Building Research Establishment – Digests and Bulletins.

7. *Other references*

(*a*) British Standards Institute – *Year Book* (British Standard Specifications and British Standards Codes of Practice).
(*b*) *Dictionary of Technical Terms.*
(*c*) *Spon's Architects' and Builders' Price Book.*
(*d*) *Construction Industry Research and Information Association (CIRIA) – Guide to Sources of Information.*
(*e*) *Steel Design Manual.*
(*f*) *Handbook on Structural Steelwork*, by British Constructional Steel Association Limited.
(*g*) *National Building Specifications.*
(*h*) *Concrete Detailing Handbook.*

4.2 Roles of the design team

The design team can be identified as those persons or firms who are invited by the client to undertake some form of work on his behalf, whether it is designing the structure or acting in the capacity of consultants to the client through the architect. They include:

1. The Architect. His assistants also play a major role in a project:
   (*a*) Chief technician.
   (*b*) Technicians/draughtspersons.
2. Quantity surveyor.
3. Structural/civil engineer.
4. Services engineers.
   (*a*) Electrical.
   (*b*) Mechanical.
   (*c*) Special services – acoustics, thermal insulation, etc.

Sales and technical representatives also give assistance in the choice of materials and components, while the local authority departments can give invaluable help on the sighting of the structure and special problems regarding fire protection, access and acceptable limits relating to the Building and other Regulations. In agreeing details in the early design period with the local authority departments time savings are made at the planning and building regulations stages.

The following brief details are given to show the extent of the work responsibilities of each member of the team:

Architect

1. Preliminary discussion with the client.
2. Discussion with client to establish the 'brief'.
3. Outline planning approval for the project.
4. Advice on appointment of quantity surveyors and other consultants.
5. Sketch plans.
6. Design drawings and other detail documents.
7. Coordination of work of all the members of the design team.
8. Full planning and building regulations consents from the local authority departments.
9. Advice on clerk of works appointment.
10. Drawing up a list of contractors for tendering.
11. Selection of contractor.
12. Superior to own technical staff.
13. Manages all affairs relating to the preparation of contract documents and work during the project duration.
14. Calls site meetings when necessary to discuss problems arising.
15. Choice of nominated subcontractors and suppliers.
16. Reports to client on progress of project.
17. Cost checks.
18. Variation orders.
19. Drawings modified if necessary.
20. Approval of interim and other certificates.
21. Final certificate issue.

Chief Technician/Draughtspeople

1. Subordinate to the architect.
2. Superior to other technicians/draughtspeople.
3. Assists in planning the work load (production of drawings and other relevant information).
4. Assists in design interpretation.
5. Detailed drawings.
6. Represents the architect when necessary.
7. Discipline of office staff.
8. Survey report.
9. Liaison with consultants.
10. Office stationery and material ordering.

Technicians

1. Subordinate to chief technician.
2. Detail drawings.
3. Land survey.
4. Specifications preparation/writing.
5. Schedules of doors, finishes, etc.
6. Structural survey when necessary.
7. Site inspection.

Professional Quantity Surveyor

1. Approximate estimates for the project.
2. Cost checks and analyses.
3. Bills of quantities, and if necessary –
   (*a*) Reduction bills.
   (*b*) Addendum bills.
4. Measurement of work on-site.
5. Interim certificates.
6. Final accounts.

Structural/Civil Engineer

1. Soil tests and analysis.
2. Advice on structural design.
3. Comparison checks on the use of steel or concrete structures.
4. Calculation of sizes of units/structural components, especially for local authority approval.
5. Structural advice throughout project duration.
6. Designing for fire protection.
7. Designing of retaining walls and other engineering structures.
8. Scheduling of reinforcement for concrete.
9. Underpinning and shoring design.

The structural engineer can be employed to undertake other specialist work in addition to that required by an architect. He/she is trained to do structural surveys, design of trench timbering or sheeting and formwork for complex concrete structural work.

Services Engineers (Internal environment engineers)

1. Lighting – artificial and natural.
2. Sound insulation.
3. Acoustics generally.
4. Thermal insulation.
5. Heating and central heating.
6. Ventilation – natural or artificial, and air conditioning.
7. Drainage.
8. Refrigeration.
9. Gas supply.
10. Water supply.
11. Telephone system.
12. Fire protection and equipment.
13. Refuse disposal.
14. Communication systems – lifts/elevators for personnel, and conveyors for materials.
15. Security system – anti-theft and pilfering devices.

Public utility services Boards can give advice and could also be employed as consultants.

If the contractor was employed at an early stage his assistance in 'techniques of construction' may lead to considerable savings being made for the client before the contract documents are finalised.

4.3 Outline plan of work by the architect

Briefing stage

Inception

Before a decision is made to build new or alter existing buildings clients must consider their needs carefully. In the first instance they must ask themselves why it is necessary to incur expenses, and usually alternative ways round the problem would be sought. However, they may find it essential to build because of the lack of accommodation and this will assist in the efficient running of their businesses. Expansion of trade also necessitates additional working space.

Private individuals may find that due to increases in their families their existing dwellings require additional bedrooms, and as the alternative of moving house is very disruptive, plans for extensions are drawn up. Other reasons for building expenditure are:

1. Change of use of premises – house into shop.
2. Conversion – house into flats.
3. Alteration – small dining and living room into one large dining/living room.

4. Renovation – improve the accommodation by repairing or renewing fittings.
5. Nuisances or dangerous state – notice served to an individual by a local authority to ensure that unsafe parts of a structure are made safe.
6. Refurbishment – updating of all, or part of, the interior or exterior of a building, i.e. services fittings, etc.
7. Speculations – purchasing land and developing it for sale.
8. Investments – one of the best ways of safeguarding capital against inflation is to invest in buildings.

Once the decision is made to build, clients can negotiate either directly with all those concerned with the future design of a structure which they have in mind, or have representatives to undertake the negotiations on their behalf. The representatives may take the form of a committee, with a chairman to steer through their recommendations and with whom the design team can liaise. This 'go-between' carries out the decisions arrived at by the committee, working party or other decision makers. With the views of the committee, etc. in mind, the client/representative next approaches an architect (see Fig. 4.3.1 for the design process).

Architect's appointment

On being approached by a client/representative preliminary discussions are held to enable the architect to make a decision on whether to accept or decline any offer made. The architect may feel that the project is beyond his capabilities, or more likely, beyond the scope of his/her organisation. A large project would require numerous technical staff, and in this case architects of small practices would have to think carefully before committing themselves to a project. The architect may also only specialise in designing specific buildings, or may be fully committed to other projects leaving little scope for accepting new commissions. Preliminary discussions before appointment is therefore essential because they give the client the opportunity to assess the architect and to see if his/her organisation is adequately structured to carry out the work.

If, after the preliminary discussions, all conditions regarding the proposed project are to the mutual satisfaction of both parties, then the appointment of the architect is made either verbally or by letter from the client. If it is a verbal appointment then the architect should verify it by writing a letter of acceptance ensuring to incorporate within it the conditions of his appointment, as laid down in the Royal Institute of British Architects' 'Conditions of Engagement', a copy of which should also be enclosed. It would be prudent of the architect to have made clear in the preliminary discussions stage the following, so that the details in the letter of acceptance are familiar to the clients:

1. Architect's fees, and at what stage payment would be expected.
2. Fees normally charged by other consultants, if required.
3. An estimate of the project's cost (very approximate).

4. The exact role of the architect, and his responsibilities.
5. Responsibility of the client – legal affairs and ownership of property; others affected by the project (neighbours).
6. With whom would the architect liaise throughout the project – addresses and telephone numbers.

The brief

On his appointment the architect should next obtain as much further information about the client's proposals (the brief) in addition to those obtained during the enquiry stage. Full consultations through additionally arranged meetings will result in the architect formulating the client's exact requirements.

The following are the minimum particulars required by an architect:

1. Time limit for the project.
2. Preliminary details of the building type, size, etc.
3. Solicitor's information regarding boundaries, covenants and other restrictions and easements.
4. Finance available.
5. Extent of the land or property, with levels if possible (ordnance survey sheets necessary).
6. Whether the land is owned by the client or is the process of purchase underway?
7. Whether outline planning consent has been or still requires to be obtained.
8. Whether there are any other agents involved, or will the architect be the sole agent?

Finally, having been given approval to use the services of consultants/specialists the client should be advised to write to them confirming their appointment, after the architect has made personal approaches to see if they are available.

Feasibility study

This process is an extension of the 'brief' and when necessary, in the light of new ideas developing or a change of mind by the client results, modifications should be made to the brief to arrive at the client's exact requirements. It is the architect's responsibility to study all aspects of the proposed project to see if the client's suggestions are feasible at this stage, and to make recommendations concerning the project in its present form.

Outline planning

If outline planning consent has not been obtained by the client's earlier consultants, such as an estate agent or solicitor, then this would be the architect's priority. So, bearing in mind the type and extent of the project, 'planning forms' would be obtained from the Technical Services Department or Planning Department of the Local Authority – an example of which is shown in Fig. 4.3.2. In obtaining 'consent to build' the client would be justified in continuing with the expense of the architect's services.

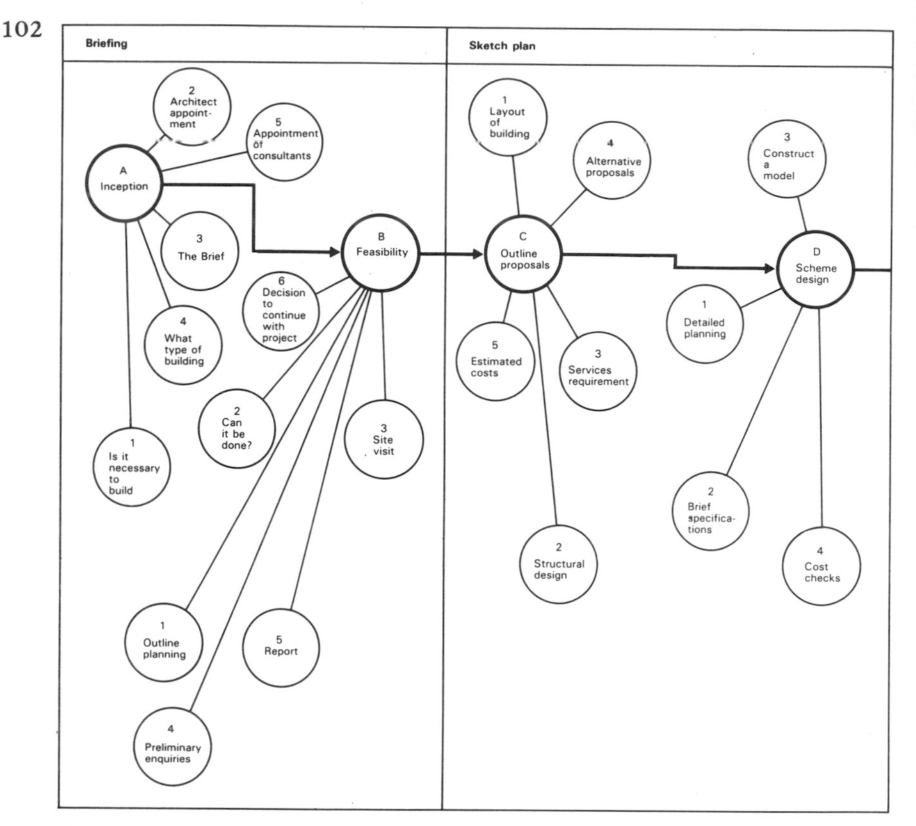

Fig. 4.3.1 Process from inception to Site Operations. (Adapted from *Plan of Work for Design Team Operations* published by RIBA Publications Ltd, London)

Site visit

The architect or his representative would visit the site as part of the 'outline planning' stage. He would record as much information as possible to assist in his quest for a satisfactory planning application, also, it would eventually help when preparing his report on his recommendations. The following should be recorded:

1. Extent of the site boundaries.
2. Conditions of the site – is it prone to flooding?; approximate state of ground; any obstacles?
3. Ground formation levels.
4. Suitable access.
5. Proximity of services – drains, water, electricity.
6. Position of adjacent buildings.
7. Types of buildings in the locality and density (buildings and persons).

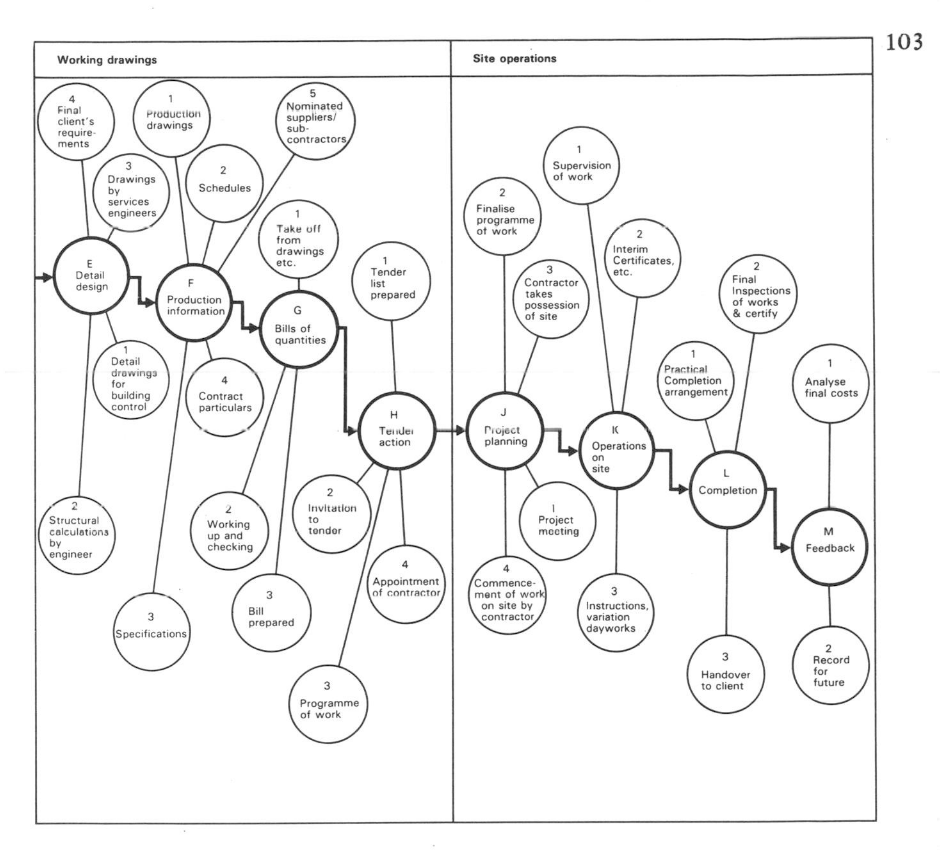

8. Information regarding similar development, alterations and extensions in the area which may give an indication of the success of outline planning applications.
9. Other features and conditions which may be relevant.

Preliminary enquiries

An approach should be made to the local/district authority's offices to discuss with their officers any problems regarding future development in the area which may affect the proposed project. Future development would be recorded in the Structure Plan or Local Plan which are prepared at district or local level. If services are not evident during the site visit then enquiries should be made to the various offices of the public utility services organisation for the position of their services' mains, and further enquiries made to the local authority offices for drains' and sewers' availability.

FOUR Completed Copies of this Form and Plans must be submitted together with **ONE** Copy of **the appropriate Certificate TO:-**

GO

THE PLANNING OFFICER
GOODFORD BOROUGH COUNCIL
Municipal Offices, 51 Low Street
GOODFORD, Branshire BA 2BY

Tel No. Goodford 66666

OFFICE USE ONLY

APPLICATION NUMBER

Date received

Date accepted

C/D

APPLICATION FOR PERMISSION OR APPROVAL TO DEVELOP LAND ETC — Part 1 All Applications

Town and Country Planning Act 1971

A/BA

PLEASE READ THE ACCOMPANYING NOTES FOR APPLICANTS BEFORE COMPLETING ANY PART OF THIS FORM

If extra space is required to answer any question please attach a separate sheet indicating the relevant question(s)

1 **Applicant**
(In block capitals)

Name ....

Address ....

....

.... Post Code ....

Tel No

Agent (if any) to whom correspondence should be sent
(In block capitals)

Name ....

Address ....

....

.... Post Code ....

Tel No

2(a) Address or location of land to which application relates ....

....

Site must be shown edged RED on the submitted site plan *See note 9* State site area in Hectares [ ] or Acres [ ]

(b) Does the site comprise land registered in accordance with Class 2 of the Community Land (Excepted Development) **Regulations 1976 as being land owned by a builder or developer on 12th September 1974?** State **Yes** or **No** [ ]

3 **Brief particulars of proposed development,** including the purpose(s) for which the land and/or buildings are to be used

....

....

....

....

4 **Particulars of application** *See notes 4 and 5*

		State **Yes** or **No**	
(a)	Outline planning permission	[ ]	If **Yes** tick any of the following which are to be considered as part of this application: 1 Siting 2 Design 3 External appearance 4 Means of access 5 Landscaping
or (b)	Full planning permission	[ ]	
or (c)	Approval of reserved matters following the grant of outline permission	[ ]	If **Yes** state the Date and Number of outline permission. Date ________ Number ________
or (d)	Continuance of use without complying with a condition subject to which planning permission has been granted	[ ]	If **Yes** state the Date and Number of previous permission and identify the particular condition. Date ________ Number ________ The condition No ________
or (e)	Permission for the retention of buildings or works constructed, or for the continuance of a use of land instituted before the date of this application	[ ]	If **Yes** state the Date when the buildings or works were constructed or carried out, or the use of the land commenced. Date ________

FORM T.P. 1 (Part 1)

Continued Overleaf

Fig. 4.3.2

5 Is the development temporary or permanent? .... If temporary state for what period ....

If a previous temporary planning permission exists, state the date and number Date Number

6(a) What is the applicants interest in the land? e.g. Owner, Prospective purchaser, Lessee, etc. ....

(b) Does the applicant own or control any adjoining land? State **Yes** or **No** [ ] If **Yes** it must be shown edged BLUE on the submitted site plan *See note 9*

7 If the application is for new residential development, state the following:-

Density in dwellings per hectare/acre	[ ]	Type (House, Flat, etc	....	Number of Garages of garage spaces	[ ]
Total number of dwellings	[ ]	Number of storeys	[ ]	Number of parking spaces	[ ]
Total number of habitable rooms	[ ]	Total gross floor area of all buildings (Sq. metres/Sq. feet)	[ ]		

See note 7

8 Does the proposed development involve:— (a) Construction of a new access to a highway? **(b) Alteration of an existing access to a highway?** **(c) The felling of any trees?**

State **Yes** or **No**	Vehicular	Pedestrian	Vehicular	Pedestrian	
If **Yes** indicate positions on plan	[ ]	[ ]	[ ]	[ ]	[ ]

(d) How will surface water be disposed of? ....

(e) How will foul sewage be dealt with?

9(a) List details of all external building materials to be used if you are submitting them at this stage *See note 10*

....

(b) List any samples that are being submitted ....

10 List all drawings, plans, certificates, documents etc; forming part of this application *See notes 9, 10, 13 and 14*

....

11(a) What is the present use of the land/buildings? If vacant, what was the last use and when did this cease?

....

(b) What buildings are to be demolished? ....

If any state gross floor area (Sq metres/Sq feet) [ ]

12(a) Are any 'Listed' buildings to be demolished? *See note 12* State **Yes** or **No** [ ]

(b) Are any 'Listed' buildings to be altered or any non listed buildings in a Conservation Are to be demolished? State **Yes** or **No** [ ]

If 'A' and 'B' is **Yes** then Listed Building Consent will be required and should be submitted on separate forms from the Planning Department

NOTE **1** Form T.P. 1 (Part 2) should now be completed for all applications involving Industrial, Office, Warehousing, Storage, or Shopping development **2** An appropriate Certificate must accompany this application unless you are seeking approval to reserved matters *(See notes 13 and 14)* **3** A separate Form is required for Building Regulation approval *(See note 16)*

PLEASE ENSURE THIS FORM IS SIGNED AND DATED BEFORE SUBMITTING

I/We hereby apply for planning permission or approval (as the case may be) for the development described herein and shown on the accompanying plans

Signed ____________________ On behalf of ____________________

Date ____________________ (Insert Applicants name if signed by an Agent)

 Special Note: The site visit survey would vary considerably depending on the type of development envisaged. An extension to existing buildings would indicate to the architect the availability of services and the suitability of the subsoil subject to the absence of cracks in the existing structure. On the other hand a completely new project on virgin ground requires a more thorough site investigation.

Consultants

These would be assisting the architect to compile suitable information to incorporate in the 'feasibility report' and would offer advice throughout.

Report

When all the information has been correlated written details in the form of a 'report' should be presented to the client for a decision on whether to continue with the project as stated in the 'brief', modify, or to discontinue because of the previously unforeseen technical problems.

If the client decides to continue with his scheme, then it now only remains for him to agree a programme for the production of information by the architect and members of the consulting team so that a construction contractor can be appointed as soon as possible.

Fees

The professional work up to this point would be charged by the consultants at an hourly rate.

4.4 Sketch plans

The procedure concerning sketch plans are as follows:

1. Outline proposals

The architect now prepares, with the assistance of all the consultants, his proposals for the approval of the client on such matters as the design and construction. The general layout of the proposed building should be decided taking into consideration the fact that the specialist consultants would be assisting in making decisions to enable the quantity surveyor to estimate approximate costs for the work. Also the decision would be influenced by the structural method proposed and the method of erection. The type of foundation construction would have a considerable influence on the costs especially if piles were necessary instead of traditional strip. The engineers could suggest one type of road in preference to another and the services necessary to give comfortable internal and external environment conditions.

The services recommended would normally take into account the capital outlay, running costs and efficiency, in addition to the ease with which the appliances and ancillary equipment fits and blends into the structure. The structural engineer would normally be consulted to check the structural design or suggest alternatives to support heavy components of the services.

Note: This presupposes that a contractor has not been appointed in the early stages of design. If one had been appointed his recommendations would be sought.

The height, size and layout of the building may be recommended by the architect to the client, but alternative proposals should be prepared to either give the client more choice, or to be used to be discounted immediately as being less favourable than the design recommended.

The estimated costs of the proposed structure is usually prepared by the professional quantity surveyor after consultations with the other members of the design team and bearing in mind the recommendations put forward by them. This would then produce a firm estimate which can only be finalised after comparing the recommendation with past projects of similar design and nature. Sketch designs would be in an advanced stage by this time as the architect would have correlated all the consultants' proposals.

2. Scheme design

It now remains for the architect and his team to prepare proper design drawings, and if necessary, a model can be prepared by the architect's model maker.

Provided that the client is still in approval with the recommended scheme the architect maintains and coordinates action on the design. 'Detailed Planning' approval is now the aim to save on possible delays in commencing work on-site later.

The full scheme design with brief specifications is prepared and further comparitive costs are calculated by the quantity surveyor. Quality and quantity of products and other units included in the drawings by the architect and engineers (structural, mechanical and electrical) would be noted by the quantity surveyor. The cost control function of the quantity surveyor ensures that throughout the design stage of a project the costs can be monitored by the architect, so that if the design is becoming too expensive, action can be taken to keep within the limits laid down by the client in his 'brief'.

In applying for 'Detailed Planning Permission' forms are obtained from the local/district offices which are usually completed in quadruplicate (see Fig. 4.3.2).

4.5 Working drawings

Detail design

The architect must now ensure that there are no more client's requirements so that proposals to date are finalised regarding all particulars relating to the design and associated costs. Detail drawings are next prepared and all members of the design team will, during this stage of the work, progress towards compiling their own detail drawings and collaborate with the architect and the other consultants where necessary. The structural engineer completes his drawings with calculations to justify the sizes of the structural components, while the service engineers prepare their drawings regarding layout of lighting circuits, heating and ventilating routes and other essential internal environmental installations. The consultants' information would be incorporated finally into the architect's detail drawings.

In coordinating the work of the design team the architect ensures that sufficient information is fed back to the quantity surveyor who continuously checks the costs against the budgeted amount.

The client should be shown the final detail drawings before they are submitted to the building controls officer for approval under the Building Regulations 1986; the minimum number of sets of drawings being two, which includes the following:

1. Key plan – scale 1:2500.
2. Block plan – scale 1:1250.
3. Each elevation – scale 1:100 or 1:200.
4. Plan of each floor level – scale 1:100.
5. Plan of roof – scale 1:100.
6. Vertical sections – scale 1:100.

Note: These scales are all minimum requirements. The building controls officers will also want proof that the sizes of structural members and thermal insulation values are adequate. It is therefore essential to provide calculations as recommended by the relevant British Standards Code of Practice. Fire protection and means of escape should be detailed carefully and the public services positions should also be shown on the drawings.

Production information

This entails the preparation of the following:

1. Production drawings.
2. Schedules.
3. Specifications.
4. Contract particulars.

Production drawings

When the detail drawings have been approved by the local authority representatives production drawings are prepared by agreement with the design team to supplement the detail drawings. Details need to be drawn larger to enable nominated suppliers and subcontractors to price their respective work easier and to produce special components which are necessary. Drawings – to scales 1:20, 1:10 or 1:5, are drawn in accordance with the 1984 BS 1192 – are therefore essential.

The additional drawings are prepared to accurately convey to the contractor the architect's requirements, and would therefore allow the main contractor to price the work more easily with the added advantage later of enabling the contractor to work without too many delays in querying details which are illustrated to a very small scale.

These are normally prepared for large jobs, and the purpose for their preparation is to simplify the presentation of information, as they can be referred to in conjunction with the production drawings. The usual types of schedules prepared by the architect – examples of which are shown in Fig. 4.5.1 – are as follows:

1. Doors schedule: to show type, quality, size, thickness and finish.
2. Windows schedule: to show details as above, including types of any glass used.
3. Internal finishes schedule: to walls, floors, ceilings, including joinery.
4. Decorations schedule: to walls, ceilings, skirtings and fittings.
5. Drainage schedule: includes gulleys, manholes (inspection chambers), rainwater pipes and other relevant parts.
6. Sanitary wares schedule: water closets, basins, sinks, etc.

There are many types of special schedules which are best prepared separately by the consultancy teams, examples being: reinforcement schedules by the structural engineer, and electrical fittings schedules by the electrical engineer.

In the first instance, the information so given in the schedules enables the professional quantity surveyor to prepare the bills of quantities more easily and more accurately. Secondly, the contractor and subcontractors would find that by using the schedules enables them to carry out certain operations quicker without the need to communicate constantly with the architect. The schedules are additional aids to the specifications, drawings and, finally, the bills of quantities.

When the schedules have been prepared in accordance with the costs yardstick, approval must be sought from the client.

Specifications

During the preparation of the production drawings the architect and his team of consultants should be preparing brief notes on the various methods of construction, components and materials, which can then be elaborated on later by the quantity surveyor when the specifications are written (assuming the architect does not prepare the specifications himself).

The reason for the preparation of specifications is to help convey to the contractor additional information about such subjects as: special problems regarding the work; standard of workmanship expected; preparations to surfaces and components; standards of materials to be used; design mixes of mortar or concrete; and to refer the contractor to the relevant BSS and BS Codes of Practice. The specifications therefore deals with the standard expected from the contractor.

Door/frame schedule																	Schedule																
Doors							Frames				Ironmongery																						
											Butts, Hinges, Springs.						Locks, Latches, Bolts.							Furniture					Stops, hooks, etc.				Letters
No.	Position	Type	Quality	Finish	Fire Rating	Glazing	Type	Finish	Sidelight	Fanlight	Size: Butts No. pr per leaf			Over/head closers	Floor springs		Size: Mortice lock					Size: Flush bolts	Cylin. rim night latch	Lever handle set	Pull handle		Kicking plate and finger plates	Door holder	Door stops	Hat and coat hook			Nos. etc.

Finishes sc					Decoration and colour			
Position	Floors	Walls	Ceilings	Skirtings	Wall	Ceilings	Skirtings	Fittings

Sanitary wares schedule						
Position	W.C.s	Lav. basins	Sinks	Shower units	Urinals	Fittings

Drainage schedule				
Position	Gulleys	M.H.	R.W.P.s Outlets and vent pipes	Channels and outlets

Fig. 4.5.1 Examples of schedule layout

These should be prepared by the architect after consulting the client, and then by agreement with the quantity surveyor and other consultants. In preparing the particulars the following should be included:

1. Names and addresses of the client, architect and quantity surveyor.
2. Description of site – position, access, working space.
3. Description of works – type of building, contents and order of completion.
4. Particulars regarding type of contract, i.e. Standard Form of Building Contract, Private Edition with Quantities, 1980, by the Joint Contracts Tribunal. The clauses dealing with dates, periods and costs are particularly discussed and decisions are made regarding the following:

    (*a*) Prime costs and provisional sums.
    (*b*) Defects liability periods.
    (*c*) Periods of interim certificates.
    (*d*) Retention limitations.
    (*e*) Insurance of the works.
    (*f*) Liquidated and ascertained damages, etc.

The decisions taken on these points would then be incorporated in the 'Preliminaries' section of the bills of quantities.

Finally at the 'Production of Information' stage it must be decided which parts of the work are to be executed by nominated subcontractors, and what materials or components should be provided by nominated suppliers/manufacturers. Invitations to tender are sent to those firms which are included on the lists drawn up by the architect, with the client's approval, and in consultation with the other consultants. If the firms so chosen are interested in tendering, all relevant information should be sent immediately to them, and the lowest bid which is received wins the nominated subcontractor or supplier's role. Obviously there are other factors to consider in addition to the lowest bid, for example: delivery dates, discounts, attendance requirements, past performances, suitability of the organisation to undertake the work/supply, to mention but a few.

Bills of quantities
(see also Chapters 8 and 9)

The professional quantity surveyor should be furnished with the production drawings, schedules (from the architect, structural engineer, service engineers), specifications, contract particulars, nominated subcontractors' and suppliers' quotations, and other relevant information to enable him/her to prepare the Bills.

Note: To assist the quantity surveyor in his preparations and to help standardise the Bills, a document known as the Standard Method of Measurement of Building Works (SMM) was compiled by the Royal Institute of Chartered Surveyors and the National Federation of Building Trade Employers in 1922. It has since been amended on numerous occasions to meet the changes which have taken place in the industry.

All items of work and quantities are calculated from the production drawings, in conjunction with the other documents, by the quantity surveyor and are listed in the bills of quantities so that the invited tenderers, in the 'selective tendering' situation, can price the work from the same information which shows exactly the extent of the work to be undertaken. The Bills, when prepared, also serves as a control document during the course of the contract, particularly when they have been priced.

An alternative to writing a separate specification document is to incorporate the specifications within the bills of quantities, thereby giving the quantity surveyor an increased reponsibility to the client.

The pages in a bills of quantities are divided into columns (see Fig. 8.4.5) and the descriptions of the work are printed clearly to enable the contractor to price every item. Each priced page is totalled and carried to a 'collection', and all the collections relating to a section of the work is carried to a 'summary'. Finally, all the summaries are collected at the back of the bills of quantities under a heading of 'General Summary'.

Clerk of Works/Resident or Visiting Engineer

This person should be appointed sufficiently early enough to be able to spend some time familiarising himself with the building contract documents. He is appointed directly by the client usually with the approval of the architect. It is therefore early enough when all the production information has been prepared.

Tender action

If contractors are to be selected a list is prepared and invitations to tender are sent out, and subject to favourable replies being received some of the tender documents are posted to the individual contractors whilst the others remain available in the architect's office for inspection by the contractors if they so wish. These documents are:

Sent by post:

1. A Tender Form.
2. General Arrangement Drawing.
3. Bills of Quantities and Specifications (see Chapters 8 and 9).

Retained in office:

4. Standard Form of Building Contract (type depends on the contract).
5. Schedules.
6. Detailed Drawings.
7. Any other information which may be relevant to assist the contractors in pricing the work accurately.

The contractor, on receipt of the contract documents, would begin the complex process of estimating the costs to do the construction work. Then by adding an additional percentage for overheads and profit would arrive at a tender figure. This is submitted to the architect as a bid for the work in competition with other contractors. The contractor who submits the lowest bid is generally awarded the contract. It is also a requirement under the JCT Standard Form of Building Contract for the bidders to submit a proposed contract programme (bar chart) to show the planned duration of the work.

4.6 Site operations

Project planning

The architect should now check that the contractor's insurances are in order for the proposed project once the contractor has been appointed. The additional production information would next be passed over to the successful contractor to allow him/her to plan the work and make final arrangements with subcontractors and suppliers and to notify the appropriate authorities (Local or Statutory Authorities).

A project meeting should be arranged to discuss all relevant points previously described and to determine an understanding with the main contractor, etc. regarding communications and the next stage of 'operations on site.' Permission should be given for the contractor to take possession of the site and to commence work.

Operations on site

Regular project meetings should be arranged by the architect to ensure control of the operations and progress by the contractor on site. It is also the responsibility of the architect, preferably through a clerk of works, to supervise each stage of the work and to ensure quality is maintained.

At each stage of the site work it is generally the duty of the architect to see that the professional quantity surveyor carries out valuations of each completed stage of the work, whereupon interim certificates are prepared so that stage payments may be made to the contractor and subcontractor to pay for work done to date – usually with a percentage retention to pay for faulty work which may develop or be subsequently discovered later.

Completion

The architect should prepare for the 'practical completion' inspection of the work done by the contractor two or three weeks before completion of the project work takes place. After the inspection a practical completion certificate should be issued to the contractor, and there should be a handover meeting, always assuming that the client had been advised to take out insurance for the structure in time for taking possession.

An inspection would be undertaken at the end of the defects liability period whereon a 'Certificate of Completion of Making Good Defects' would be issued. Ultimately final accounts are prepared to arrive at the sum still owed to the contractor.

Feedback

Lastly, all the final costs should be analysed and records should be made for future reference.

Chapter 5

The town planning situation

5.1 Historical development and control of development

Historical

In the first instance man planned to be close to his sources of nourishment and this meant living in close proximity to rivers or streams which supplied the means of life. Shelter was also necessary to give protection from the weather and predators. Through his search for food he also found that satisfactory clothing could be derived from the skins of animals as additional protection from the elements.

Groups later formed for the mutual benefit of all, and areas were reserved for meetings of councils, and plots of land were made available to the children which ultimately developed into settlements. Areas which were selected for settlements normally gave the additional benefits of good defences from possible enemies. The strong, however, began to take advantage over the weak, and organised armed bands or tribes conquered others and later formed larger civilised communities.

With the advent of large armies empires were formed due to the conquest of people in distant lands; the Assyrians, Babylonians, Egyptians, Greeks and later the Romans being the most successful conquerors of their time.

The search for better standards of living meant that colonies were created, and to give protection from the existing hostile populations fortifications were built. Villages, towns and cities eventually grew close to these fortifications and trade began to flourish because there were large concentrations of people, such as the armies and their followers.

The most successful town planning was undertaken by the Romans, although many other civilisations planned their towns well. The grid system of town layout was used and took into consideration the provision of shops, churches and other public meeting places. Cemeteries were usually excluded from within the town or city.

Flourishing civilised nations in ancient times provided prestige structures – the Mesopotamians with their ziggurats; the Egyptians with their pyramids and obelisks; the Greeks with their temples, mausoleums, circuses and odeons and the Romans with their amphitheatres, triumphal arches and other monuments. This type of prestige development still continues, as can be seen throughout the countries of the world.

The Normans introduced a system which gave ownership of all the land to the King and in turn it was leased to the Barons for monetary and other favours. It was then the responsibility of the Barons to sub-let the land, thereby reaping a percentage of the produce of the land from the tenants. Social, economic military and religious problems were partially solved within this system.

In the Middle Ages due to the growth in the population, particularly in a town such as London, the poor layout of the dwellings and the lack of proper public health precautions led to many plagues, resulting in the death of a large percentage of the population. Haphazard development throughout Britain led to conditions which were most unsatisfactory for the inhabitants. With the outbreak of the Plague in 1665 and the eventual catastrophe of the Great Fire of London a year later, the authorities began to reappraise the layout of their city.

When the City was rebuilt after the appalling losses to life and property, a new Act of Parliament laid down rules for the rebuilding programme and architects were appointed for the design and layout. Town planners, as such, did not exist at that time and the work was normally carried out by architects. While this led to the earliest form of Building Regulations it did not significantly affect town planning.

It was not until the Industrial Revolution of the mid-seventeenth century to mid-eighteenth century that the problem of lack of control over development by the authorities began to be realised. Too many industrialists and speculators controlled the way land was used or abused. Knowingly or unknowingly haphazard development took place with the one criteria being to provide accommodation for employees as close to their places of work as possible for convenience and cheapness. Very little regard was placed on good layout, sanitation or the environment generally. Squalid conditions resulted in many districts and epidemic diseases were not infrequent.

The middle and upper classes had been fortunate in living in areas which were less populated, and their dwellings were proudly designed by such masters as Inigo Jones, Christopher Wren, John Nash and others.

New towns were built and existing villages grew to accommodate workers where mineral discoveries were made. Urbanisation developed to large proportions with the drift of agricultural workers to new employment within the factories which were increasingly being built. Different styles of communication systems grew in addition to the improved roads, such as canals and railways. The use of steam, gas and later electricity added eventually to the growth of the urban areas. The wider use of these so-called improvements only added to the problems of overcrowding and misuse of resources leading to very ugly and sometimes unhygienic conditions for the population.

As conditions worsened attempts were made to control them which led to a series of Public Health Acts being passed by Parliament commencing in 1848, and culminating for the time being in the Public Health Act 1875 which consolidated all previous Acts. This formed the basis of a constructive start to a whole range of legislation affecting health, layout of dwellings and town planning as we know it today.

The Working Classes Dwellings Act 1890 allowed for the clearance of slums which then enabled properly constructed dwellings to be erected by local authorities. Many Acts followed dealing with public health until the Public Health Act 1936 (which is still the principal Act) was introduced, and later was amended by the Public Health Act 1961. There are other Acts which are concerned with public health, firstly, the Clean Air Acts 1956 and 1968, and secondly, the Thermal Insulation (Industrial Buildings) Act 1957 followed by the Thermal Insulation (Industrial Buildings) Regulations 1972.

The earliest attempt at town and country planning was made with the introduction of the Housing, Town Planning, etc. Act 1909 by the Government. Ebenezer Howard had foreseen the necessity of building Garden Cities, the first of which was commenced at Letchworth in 1904. This type of city provided an overspill from the large cities which were becoming overcrowded, thereby leading to poor social and environmental conditions. The concept of Garden Cities was to provide a self-sufficient and agreeable environment for the inhabitants.

The 1909 Act eventually led to the realisation that town planning was essentially a separate problem, and later was treated as such by the introduction of legislation additional to housing and public health. The advent of this new Act was primarily to allow local authorities to prepare schemes for the future development of parcels of land, and to control undesirable development. Any schemes envisaged by a local authority had to have the prior approval of the Local Government Board. This system was later mainly dispensed with under the Housing Town Planning Act, etc. 1919; however, schemes when prepared still had to be approved by the Local Government Board. The Act also stipulated that certain local authorities had to prepare schemes, which was the earliest attempt at regional planning leading later to the principle of 'development plans'.

The Town Planning Act 1925 finally heralded the separation of town planning from housing, and with the passing of the Local Government Act of 1929 county councils were given the responsibility of preparing development schemes, especially when district councils failed in their responsibility to do so when directed by the Local Government Board.

The Town and Country Planning Act 1932 repealed the other town planning legislation but incorporated some of the previous Acts' sections, but with modifications. In addition, the need to obtain permission from the Minister to prepare a planning scheme was necessary, and when prepared it later had to be approved by Parliament. Local Authorities under the Act were now empowered to make schemes affecting land which was already built upon. As the 1932 Act contained the word 'Country', it additionally related to the preservation of trees and control of minor development, such as hoardings and advertisements. This Act contained many more sections than was incorporated in the town planning Acts before it, thereby showing the importance being placed on matters relating to proper development and conservation of the environment.

In 1943, two Acts relevant to planning were introduced; one created the position of Minister of Town and Country Planning, thereby relieving the Minister of Health from town and country planning responsibilities. The Town and Country Planning Act 1943 lasted until 1951. The second Act was the Town and Country Planning (Interim Development) Act 1943, which required any developer to obtain permission from the local authority before development commenced. Failure to comply could result in the loss of compensation on removal of the disapproved structure by the local authority. Under the old system contained in the 1919 Act development could have continued under the 'interim development' scheme, and if permission was then refused, the developer could claim compensation for the work already done.

During the Second World War extensive damage resulted from enemy bombing of towns and cities, so the Town and Country Planning Act 1944 gave powers to the local authorities to compulsorily purchase areas of the blitzed land, and any blighted areas resulting from poor layout and obsolete development. This Act gave some responsibility to the local authorities to undertake development within their jurisdiction.

Because many changes had now taken place relating to town and country planning matters, the Town and Country Planning Act 1957 was introduced which repealed all previous ones but immediately re-enacted some past sections contained particularly in the 1944 Act. The new Act had incorporated within it a number of extremely important parts, two of which are: the number of planning authorities responsible for the preparation of development plans were reduced and the numbers had to be reviewed every five years; owners of land, in the main, had to obtain planning permission to develop their land or change its use. Other points related to tree preservation orders; orders for the preservation of buildings of special architectural or historical interest, listed buildings and control of advertisements. Owners were entitled to enjoy the use of their land in its present form, but the State's permission had to be obtained to develop it in any way. Also, the new Central Land Board under the Act were empowered to levy development charges. This gave the interests in the development value of land to the State and not to the owner.

The Town and Country Planning (Amendment) Act 1951 and the other Town and Country Planning Acts 1953, 1954 and 1959 amended parts of the Principal Act of 1947, the main alteration vesting the development value in favour of the owner.

The Town and Country Planning Act 1962 repealed the previous Acts from 1947 onwards and remained the principal Act until 1971. It was amended in 1963, 1965, 1966 and 1968, and further Acts helped to control development such as the Control of Office and Industrial Development Act 1965 and the Industrial Development Act 1966. The 1963 Act restricted the enlargement of existing buildings without planning permission. Similarly, with the introduction of the Control of Office and Industrial Development Acts, office 'permits' were required for new, or extensions to existing, office buildings, while industrial development 'certificates' were necessary for industrial buildings depending on the size and location regarding both types of development. Both the permits and certificates were essentially required from the Minister before planning applications were made to the local authority although the requirements have since been repealed regarding permits. A new system was introduced in the Town and Country Planning Act 1968 which divided development plans into two parts. These were Structure and local plans which are now essential features of planning, and the two types of plans, when prepared by a local planning authority, help to control all development within an area. With the advent of the Town and Country Planning Act 1971, at present the principal Act, the two-level system still prevails. The Town and Country Planning (Amendment) Act 1972 and the Town and Country Planning Amenities Act 1976 makes further provisions, particularly relating to the control of development which affects amenities enjoyed by the public.

Acts of Parliament

The Acts which are relevant to planning in England and Wales, for which similar Acts exist in Scotland and Northern Ireland, are as follows:

Ancient Monuments Acts 1913 and 1931.
Agriculture Act 1970.
Agriculture (Miscellaneous Provisions) Act 1963.
Coast Protection Act 1949.
Community Land Act 1975.
Commons Act 1976 and 1899.
Commons Registration Act 1965.
Compulsory Purchase Act 1965.
Control of Pollution Act 1974.
Countryside Act 1968.
Caravan Sites Act 1968.
Caravan Sites and Control of Development Act 1960.
Civic Amenities Act 1967.
Civil Aviation Act 1971.
Clean Air Acts 1956 and 1968.

Clean Rivers (Estuaries and Tidal Waters) Act 1960.
Deposit of Poisonous Waste Act 1972.
Factories Act 1961.
Forestry Act 1967.
Green Belt (London and Home Counties) Act 1938.
Housing Acts 1957, 1969 and 1980.
Housing Amendment Act 1973.
Health and Safety at Work Act 1974.
Housing Subsidies Act 1967.
Housing Finances Act 1972.
Housing Rents and Subsidies Act 1975.
Historic Buildings and Ancient Monument Act 1953.
Highways Acts 1980.
Industry Act 1972.
Industrial Development Act 1966.
Land Compensation Act 1961.
Land Compensation Act 1973.
Local Authorities (Historic Buildings) Act 1962.
Litter Act 1958.
Local Authorities (Land) Act 1963.
Local Government Acts 1960, 1962 and 1974.
Metropolitan Commons Act 1866.
Mines and Quarries Act 1954.
Mineral Workings Acts 1951 and 1971.
New Towns Acts 1946, 1965, 1968, 1971 and 1981.
National Parks and Access to the Countryside Act 1949.
Noise Abatement Act 1960.
Office, Shops and Railway Premises Act 1963.
Opencast Coal Act 1958.
Public Health Acts 1936 and 1961.
Pipe Lines Act 1962.
Rivers (Prevention of Pollution) Acts 1951 and 1961.
Rural Water Supplies and Sewerage Acts 1944 and 1961.
Road Traffic Acts 1960 and 1962.
Town and Country Planning Acts 1947, 1959, 1962, 1968 and 1971.
Town and Country Planning (Amendment) Act 1972.
Town and Country Planning (Minerals) Act 1981.
Town and Country Amenities Act 1974.
Town Development Act 1952.
Trees Act 1970.
Water Resources Acts 1963 and 1968.
Water Acts 1945, 1948 and 1973.

On the passing of an Act of Parliament there usually follows a number of Regulations, Rules, Notices and Orders which are prepared by the person (usually a Minister) with the delegated responsibility of carefully controlling the requirements laid down in the appropriate Act.

5.2 Structure and local plans

A local planning authority is charged with the responsibility for the preparation of development plans to show clearly its intentions on how the area under its control will be developed and used in the future, taking into consideration conservation, improvements and development of land. The social, economic, physical and environmental aspects of an area are drawn up by the authority and due regard is paid to general improvements for the inhabitants of the area.

Planning area survey

A survey of the existing features of an area is essential, and must be conducted by the local planning authority to assess present conditions, thereby enabling proposals for improvements to be made in the development plans which follow under the Town and Country Planning Act 1971. The proposals will then be up-dated periodically as and when conditions dictate. The survey would relate primarily to the following:

1. The use of all the land.
2. Size of the population, its density and distribution.
3. Industrial and commercial development.
4. Transport and other communication systems.
5. The effects past development have had on the area.

Joint surveys of a particular area may be necessary by two or more authorities if the land affects each, as in the case of National Parks which may adjoin two or more authorities' areas.

On completion of the survey development plans are prepared by the local planning authority, and every opportunity must be given for the general public to participate and contribute to the final plans which are next submitted for approval to the Secretary of State. Advertisements in newspapers, posters and public displays should be used to inform the public of the authority's proposals and to encourage them to voice their opinions, and show their approval or state objections of the schemes put forward.

Development plans have to be prepared to convey to everyone, including the Government, the local planning authority's intentions of what development is proposed for the future regarding in particular the uses of areas of land within the authority's jurisdiction. A development plan is divided into two types:

1. Structure plan.
2. Local plan.

Structure plan

This is prepared on completion of the survey of an area by the local planning authority and after the public have been fully consulted. A copy is next sent to the Secretary of State who will either approve it or insist on it being amended.

The plan is a written statement and must indicate the authority's intentions for the area over the forthcoming twenty years or more. National and regional planning policies must be observed and the structure plan will need to be prepared with the policies in mind. Any major changes envisaged are highlighted in the plan, and the local planning authority must show how the decisions were arrived at, and how and in what phases the proposed development will take place. Diagrams should reinforce the proposals and these would show the approximate plan layout of an area (not scale drawings) to outline the authority's intentions.

When the Secretary of State finally approves the structure plan or plans they then become known as the 'operative structure plans', and can be inspected at any reasonable time during normal working hours at the local planning authority's offices.

Areas on the structure plan which are to be extensively developed are known as 'action areas', and the authorities must begin the development improvement/redevelopment within ten years of the plan being approved, as stated in the Town and Country Planning Act 1971.

Local plan

On approval of the structure plan the local planning authority must prepare local plans for any action areas shown on the structure plan, these are known as 'action area plans'. Local plans are divided into three types:

1. Action area plans – proposals for intensive changes, major development, redevelopment inprovements or a combination of two or more of these.
2. District plans – proposals for comprehensive planning of large areas, such as a town, where the development is to take place in a piecemeal fashion over a long period.
3. Subject plans – detailed treatment of special parts of a structure plan such as derelict land and motorway routes.

Local plans consist of written details and a map, usually of the ordnance survey type, which must adhere to the principles outlined in the structure plan. The information so conveyed will detail precisely the changes which will be made to an area indicating the individual properties which will be affected. The normal rules for allowing the public participation in formulating and approving the plans still stands, but the plans do not require the Secretary of State's approval as they are an elaboration on the structure plans.

Note: With the existence of the development plans the planning authority uses them as a basis for making decisions on whether to permit or reject planning applications. If an area is designated for a particular type of development and an application which is incompatible is made by a developer, then the planners have grounds for refusal.

Insufficient research into the planning situation by a developer could lead to unnecessary expenses being incurred by employing different professional practitioners in the negotiations for the purchase of land, and for the preparation of designs for a particular structure. Inspection of the development plans at the local authority's offices or an approach to the local planning officer will reveal the future proposed development, and abortive time and effort can be avoided.

5.3 Types of development, certificates and notices

Development

This relates to an almost unlimited number of areas and fields of activity which are controlled by the Town and Country Planning and associated Acts. These areas, etc. need to be controlled to prevent unwanted and unsightly development, and to protect the amenities for the general enjoyment and well-being of all. The types of development which can be made and which are subjected to planning control can be broadly divided as follows:

1. Development regarding dwellings.
2. Industrial development.
3. Office development.
4. Outdoor advertising.
5. Preservation of ancient or historical buildings or monuments.
6. Development relating to new towns.
7. General conservation of the countryside.
8. Preservation of trees, etc.
9. Highways – parking, roads, bridges, footpaths, bridleways.
10. Other development undertaken by the national or local Governments and statutory undertakings.

Planning control over development is far reaching and developers and industrialists, including individuals, need to bear in mind the relevant Acts which safeguard against improper development.

Certificates, etc.

Industrial Development Certificates

When making a planning application to the local planning authority for industrial development it is necessary first to obtain an 'Industrial Development Certificate' from the Department of Trade and Industry which should then be submitted in addition to the normal planning application forms and certificates.

The types of industrial development requiring a Certificate are as follows:

1. The erection of industrial floor space exceeding 4656 m^2.
2. Extensions to existing industrial buildings which will be in excess of 4645 m^2.
3. Change of use of a building not previously used for industrial purposes into industrial accommodation.

The purpose behind the certificate is to control industrial development by directing it to areas where there is high unemployment. The certificate in itself does not automatically allow development to commence, only the local planning authority can give this approval with the structure or local plan in mind.

Ownership Certificate

A local planning authority will not entertain a planning application unless accompanied by an ownership certificate. There are four certificates each of which is obtainable from the planning department or technical services department of the local authority, and should be signed by the applicant depending on which conditions prevail (see Figs. 5.3.1, 5.3.2 and 5.3.3). They are as follows:

1. Certificate 'A' – Prepared by the applicant having a freehold or leasehold interest in the land with an unexpired duration of at least seven years.
2. Certificate 'B' – Prepared by the applicant who is part freehold owner or prospective purchaser of the land, if all the owners of the land are known.
3. Certificate 'C' – Prepared by the applicant who is part freehold owner or prospective purchaser of the land, but can only ascertain some of the owners.
4. Certificate 'D' – Prepared by the applicant who is part freehold owner but who is unable to ascertain the other owners of the land.

Notices

There are two Notices which are numbered 1 and 2, and are obtainable from the local planning authority offices when conditions dictate the use of Certificate B, C or D.

Notice No. 1: This is served on other persons with a freehold or leasehold interest in the land, if they are known, an example of which is shown in Fig. 5.3.4.

Notice No. 2: This is prepared for publication in the local newspaper where persons with a freehold or leasehold interest in the land are unknown.

Local authority certificates, etc.

Listed Building Consent and Building Preservation Notices

Buildings of architectural or historical significance should be placed on a List, the preparation of which is vested in the Secretary of State under the Town and Country Planning Act 1971. Copies of the List are retained by the local planning authority in an effort to safeguard the Country's heritage. Owners of such listed buildings should be notified when their buildings have been selected for incorporation on the List. Any demolition, extensions, alterations or modifications by the owner would then require listed building consent from the local planning authority. In addition, notification must be given to the Royal Commission on Historical Monuments (England), or a similar body in other parts of the United Kingdom. A local planning authority can serve a Building Preservation Notice on individuals when a building of special

TOWN AND COUNTRY PLANNING ACT 1971

CERTIFICATE UNDER SECTION 27

COMPLETE ONE ONLY OF THE FOLLOWING CERTIFICATES AND STRIKE THROUGH THOSE WHICH DO NOT APPLY

CERTIFICATE A **For Freehold Owner (or his/her Agent)**

I hereby certify that:-

(a) 'owner' means a person having a freehold interest or a leasehold interest the unexpired term of which was not less than 7 years

1. No person other than the applicant was an owner (a) of any part of the land to which the application relates at the beginning of the period of 20 days before the date of the accompanying application.

2. ***Either** (i) None of the land to which the application relates constitutes or forms part of an agricultural holding;

***or** (ii) *(I have) (the applicant has) given the requisite notice to every person other than *(myself) (himself) (herself) who, 20 days before the date of the application, was a tenant of any agricultural holding any part of which was comprised in the land to which the application relates, viz:-

(b) If you are the sole agricultural tenant enter 'None'

Name of tenant (b): Address

Date of Service of Notice: Signed ....

*On behalf of ....

*Delete where inappropriate

Date ....

CERTIFICATE B **For Part Freehold Owner or Prospective Purchaser (or his/her Agent) able to ascertain all the owners of the land.**

I hereby certify that:-

(#) See Note (a) to Certificate A

1. *(I have) (The applicant has) given the requisite notice to all persons other than *(myself) (the applicant) who, 20 days before the date of the accompanying application were owners (#) of any part of the land to which the application relates, viz:-

Name of owner: Address:

Date of Service of Notice:

2 ***Either** (i) None of the land to which the application relates constitutes or forms part of an agricultural holding;

***or** (ii) *(I have) (the applicant has) given the requisite notice to every person other than *(myself) (himself) (herself) who, 20 days before the date of the application, was a tenant of any agricultural holding any part of which was comprised in the land to which the application relates, viz:-

(b) If you are the sole agricultural tenant enter 'None'

Name of tenant (b): Address:

Date of Service of Notice: Signed ....

*On behalf of ....

*Delete where inappropriate

Date ....

Note: If the applicant is a tenant and does not hold a lease the unexpired term of which is less than 7 years Certificate B should be completed and Notice served on the landlord.

Fig. 5.3.1

CERTIFICATE C **For Part Freehold Owner or Prospective Purchaser (or his/her Agent) able to ascertain some of the owners of the land**

I hereby certify that:-

(a) Insert date of application

1. (i) *(I am) (The applicant is) unable to issue a certificate in accordance with either paragraph (a) or (b) or Section 27(1) of the Act in respect of the accompanying application dated (a)....................................

(#) See Note (a) to Certificate A

(ii) *(I have) (The applicant has) given the requisite notice to the following persons other than *(myself) (the applicant) who, 20 days before the date of the application were owners (#) of any part of the land to which the application relates, viz:-

Name of owner: Address:

Date of Service of Notice

(iii) *(I have) (The applicant has) taken the steps listed below, being steps reasonably open to *(me) (him) (her) to ascertain the names and addresses of the other owners of the land or part thereof and (have) (has) been unable to do so:

(b) Insert description of steps taken

(b)..

....

....

(iv) Notice of the application as set out below has been published in the:

(c) Insert name of local newspaper circulating in the locality in which the land is situated

(d) Insert date of publication (which must not be earlier than 20 days before the application)

(c)..

on (d)...

(Attach copy of notice as published)

2. ***Either** (i) None of the land to which the application relates constitutes or forms part of an agricultural holding;

***or** (ii) *(I have) (the applicant has) given the requisite notice to every person other than *(myself) (himself) (herself) who, 20 days before the date of the application, was a tenant of any agricultural holding any part of which was comprised in the land to which the application relates, viz:-

Name of tenant (e): Address

Date of Service of Notice Signed.................................

*On behalf of.................................

(e) If you are the sole agricultural tenant enter 'None'

*Delete where inappropriate

Date.................................

see seperate sheet for Certificate D and forms of Notice

Fig. 5.3.2

CERTIFICATE D — **For Part Freehold Owner or Prospective Purchaser (or his/her Agent) unable to ascertain any owners of the land other than Himself/Herself.**

I hereby certify that:-

1. (i) *(I am) (The applicant is) unable to issue a certificate in accordance with Section 27(1)(a) of the Act in respect of the accompanying application dated: (a) ....

(a) Insert date of application

(#) 'owner' means a person having a freehold interest or a leasehold interest the unexpired term of which was not less than 7 years

and *(have) (has) taken the steps listed below, being steps reasonably open to *(me) (him) (her), to ascertain the names and addresses of all the persons, other than *(myself) (himself) (herself) who, 20 days before the date of the application were owners (#) of any part of the land to which the application relates and *(have) (has) been unable to do so:

(b) Insert description of steps taken

(b) ....

....

....

(ii) Notice of the application as set out below has been published in the:

(c) Insert name of local newspaper circulating in the locality in which the land is situated

(d) Insert date of publication (which must not be earlier than 20 days before the application)

(c) ....

on (d) ....

(Attach copy of notice as published)

2.* **Either** (i) None of the land to which the application relates constitutes or forms part of an agricultural holding;

***or** (ii) *(I have) (the applicant has) given the requisite notice to every person other than *(myself) (himself) (herself) who, 20 days before the date of the application, was a tenant of any agricultural holding any part of which was comprised in the land to which the application relates, viz:-

(e) If you are the sole agricultural tenant enter 'None'

Name of tenant (e): Address:

Date of Service of Notice

Signed ....

On behalf of ....

Date ....

*Delete where inappropriate

see separate sheet for Certificates A, B and C

Fig. 5.3.3

NOTICE NO 1
For Service on Individuals

Town and Country Planning Act 1971
Notice under Section 27 of application for planning permission
Proposed development at *(Insert address of location of proposed development)*

....

....

TAKE NOTICE THAT application is being made to the Seeside Borough Council

by *(insert name of Applicant)*

for planning permission to *(insert description of proposed development)*

....

....

....

....

....

....

If you should wish to make representations about the application you should make them in writing within 20 days of the date of service of this notice, to the Planning Officer, Seeside Borough Council, Council Offices, High Street, Seeside, Sussex SE2 3XY.

Signed on behalf of
(insert applicants name if signed by agent)

Date

NOTICE NO 2
For publication in Local Newspaper

Town and Country Planning Act 1971
Notice under Section 27 of application for planning permission
Proposed development at *(insert address of location of proposed development)*

....

....

NOTICE IS HEREBY GIVEN THAT application is bėing made to the Seeside Borough Council

by *(insert name of applicant)*

for planning permission to *(insert description of proposed development)*

....

....

....

....

....

....

Any owner of the land (namely a freeholder or a person entitled to an unexpired term of at least 7 years under a lease) who wishes to make representations to the above mentioned Council about the application should make them in writing within 20 days of the date of publication of this notice to the Planning Officer, Seeside Borough Council, Council Offices, High Street, Seeside, Sussex SE2 3XY

Signed on behalf of
(insert applicants name if signed by an agent)

Date

Fig. 5.3.4

interest is in danger of being demolished, and at the same time it would apply to the Secretary of State to have the building included on the List.

Tree Preservation Order

Anyone wishing to fell trees must check if the particular trees are covered by a Preservation Order. Local authorities have the power to make tree preservation orders to prevent or control the felling of trees. Where a Preservation Order exists, a tree-felling licence must be obtained when felling is necessary.

All trees in a Conservation Area are protected, and normally where the trunks of trees in other areas are greater than 75 mm diameter a licence is required before felling.

Completion Notice

Because it is sometimes financially expedient to do so, some developers delay the completion time on a project so that for the immediate foreseeable future it is doubtful if the development will be completed. Where a local authority think that this is the case, they will serve on the developer a Completion Notice which becomes effective after 12 months. The type of development to which this relates is where planning permission has been given but the work has not been completed within 5 years, nor does it look like being completed within a reasonable time.

Enforcement Notice

This will be served by a local planning authority on those who have an interest in any development which has not had planning consent. If the authority also wish the development to cease immediately it can serve a Stop Notice on the same individuals because the Enforcement Notice does not, under the 1971 Act, come into force until the expiration of 28 days.

'Established-Use' Certificate

Where persons have an interest in land which has been developed since 1963 without planning consent, they can apply for an 'Established–Use' Certificate to certify that the development, because of the passing of time and the absence of objections to the development in its present form, is established.

5.4 Planning applications

In making an application for planning permission one must take cognisance of the details laid down in the Town and Country Planning Act 1971, and amended in 1972. An application is made on a Part 1 Form (see Fig. 4.3.2), but if industrial or office development is to be undertaken above the limits laid down, then a Part 2 Form must also be prepared which is specifically for this purpose. The forms are obtainable from the local planning authority's offices.

Certain development is permitted within the meaning of the 1971 Act, and provided that a careful check is made work can commence immediately.

 There is, however, the Building Regulations which need to be satisfied for new construction work, extensions and most alterations. Some types of development require no planning permission (exemptions), examples of which are outlined as follows:

1. Increasing the size of an existing terrace house by not more than 50 m^3 or 10 per cent of its cubic capacity up to a maximum of 115 m^3, whichever is the greater.
2. Increasing other original houses by not more than 70 m^3 or 15 per cent of their cubic capacity up to a maximum of 115 m^3, whichever is the greater.
3. Erection of walls, fences, etc. not exceeding 2 metres high (nor more than 1 metre high adjoining a highway used by vehicles. Check in the Town and Country Planning General Development Order 1977 and amendments 1980 and 1981.

Prospective developers do not have to own the land to which planning permission is sought. If the developers wish to obtain planning approval in principle before purchasing land, he/she may do so and this will safeguard against unnecessary expenditure should the planning applications be refused.

For the proposed development one of two applications can be made, which are:

1. Outline Planning Application.
2. Detailed Planning Application.

Outline Planning Permission

Either the owner of land who wishes to sell at the maximum value or a prospective purchaser may apply for Outline Planning Permission (permission in principle). In making an application only a plan of the land layout and drawings of the type of development to be undertaken need to be furnished to the local planning authority in addition to the statutory forms. The planning authority will then check each applicant's submitted details against the structure and local plans to see if the type of development will be compatible with the county or local proposals. Approval at the Outline Planning stage is operative for three years at the end of which a Detailed Planning Application must be made.

Registration

All applications should be recorded on a Register along with the eventual planning decisions, and the Register should be made available during normal working hours to anyone wishing to make reference to it. Persons contemplating applying for planning consent would find the Register particularly useful as a guide to the rate of planning application successes in which the type of development is envisaged. It would also help in preparing a case of an Appeal where a planning application has been refused.

A Detailed Planning Application can be made and Outline Planning can be dispensed with if the land is owned by the applicant and the total development requirements are known. Alternatively, before the end of the Outline Planning Approval period (three years) a Detailed Planning Application can be submitted to the local planning authority.

In applying for Detailed Planning Permission it is normal practice – in addition to the preparation of suitable plans, drawings and specifications – to obtain from the local planning authority offices the necessary forms which should then be completed by the applicant or applicant's representative. The forms are:

1. Application for Permission to Develop Land, etc. Part 1 (and Part 2 if Industrial or Office Development is considered).
2. Appropriate Certificate A, B, C or D. If Certificate C or D is used it is a statement to the effect that an advertisement has been placed in the local newspaper.
3. Notices 1 or 2, if required.
4. Industrial Development Certificate, if and when required.
5. Fee.

Four copies of the drawings and application forms are necessary to enable the local planning authority to check the information contained within the documents. When approval is given it remains operative for five years, and if Outline Planning consent was given earlier, then the period between the Outline Planning and Detailed Planning consents should be deducted from the five years, and the remainder will be the operative period of the planning approval. If work is not commenced on-site within this period a fresh application may have to be made.

Appeals

If a decision is not given within two months of the application being submitted an Appeal can be made to the Secretary of State for the Environment. More usually, however, an Appeal is made where a planning authority has refused a planning application.

An Appeal is made within six months of the decision by the local planning authority on forms provided. The Secretary of State can refuse an Appeal, but when one is allowed an Inspector is appointed by him and an enquiry is held at which both sides will be permitted to state their cases.

Chapter 6

The building act regulations

6.1 Historical aspects and legislation

Most buildings constructed in modern times using techniques which have been evolved during the past thousand years are referred to as traditional buildings. The methods of construction found by 'trial and error' over the years have proved successful, having been tried and approved by both builders and occupants alike. Components and structural elements adopted today are the results of well-tried techniques used in different areas of the country, and due to better communications systems being introduced a gradual standardisation of construction was developed; this however was still a very slow process in the United Kingdom.

During the late twelfth and early thirteenth centuries, control over the crenellation of fortified houses resulted, and licences had to be paid for by the owners. This did not enhance in any way the health and safety of the occupants but was a form of taxation for the Government.

Timber-framed buildings with infills of either wattle and daub, lath and plaster, or brickwork were constructed in large numbers up until the mid-seventeenth century, and certain control was attempted particularly where fires flourished or areas became overcrowded.

As stated in Chapter 5, London introduced a form of building control after the unfortunate catastrophe of the Great Fire in 1666. Measures were taken to prevent a recurrence in the future by the introduction of the London Building Act 1667, which placed many restrictions on development and the ways in which buildings were to be constructed. The Act was primarily a Fire Act.

In the eighteenth century legislation insisted that with certain buildings the dividing walls between terraced structures in London were to be built above the level of the pitched roof to help prevent the spread of flames to adjoining premises. Other cities followed the examples set in London, but because the largest British city had more problems than most, it continued to pass laws governing development within its boundaries. Even today, London has its own Building Regulations which are strictly applied regarding building work.

Insanitary conditions in urban areas became acute, culminating in outbreaks of cholera and other epidemics because of the drift from the rural areas by the population to the new factories and associated industries during the nineteenth century. The new public health problem became an issue in itself and later a series of Public Health Acts were introduced, the first being in 1848 which permitted new local health boards to improve conditions, such as providing sewers, clean streets and supplies of water – but the Act did not insist on these provisions, it only permitted them. While the Act was a step in the right direction, finance was not fully available and therefore did not allow for full-scale implementation.

The Public Health Act 1875 and the eventual introduction of Model Byelaws in 1877 empowered local health boards to introduce local byelaws to control certain standards of new construction. The byelaws were only relevant to the district in which they were introduced. The Act covered many aspects regarding health and welfare and remained the Principal Act until 1936; it was amended in 1890, 1907 and 1925.

The responsibility for the enforcement of the Building Byelaws was entrusted with local authorities on the introduction of the Public Health Act 1936 and eventually, in 1953, Model Building Byelaws with the 'deemed-to-satisfy' clauses enabled building control to be maintained. These clauses allow new techniques to be introduced by developers due to changing situations brought about by research and development. British Standards Specifications and Codes of Practices are now quoted as suitable standards and have continued to be so for over forty years.

The 1953 Building Byelaws remained operative until the Public Health Act 1961 made provisions for the preparation of the Building Regulations which were eventually introduced in 1965. These became the first National Regulations controlling building construction. The Building Regulations 1965 applied to both England and Wales, but London and Scotland had separate Regulations. The 1965 Regulations were amended six times. After the advent of 'metrication' they were all consolidated in the Building Regulations 1972. Once again it was not long before even these Regulations were amended in 1973, 1974 and 1975. The Building Regulations 1976 came into force on 31 January 1977, and was followed by amendments in 1978, 1980 etc. until new Regulations were introduced in 1985.

 The power to make building regulations lies with the secretary of state while local authorities have, in the main, the responsibility to ensure compliance by individuals and organisations.

Acts of Parliament

The Government passes new legislation as conditions change regarding standards of construction. Amendments to existing legislation and the introduction of new ones helps to ensure the maintenance and improvement of the environment of us all. The following Acts contribute to the control of building work:

The Building Act 1984.
Health and Safety at Work etc Act 1974.
Offices, Shops and Railway Premises Act 1963.
Fire Precautions Act 1971.
Public Health Act 1936, 1961.
Clean Air Act 1956, 1968.
Highways Act 1980.

6.2 Building Act/Regulations: parts and schedules

The Building Act 1984 (in outline only).
The Act is divided into 5 parts as follows:

Part I – Building Regulations

This part gives the Secretary of State the power to make Regulations for the purposes of:

(a) securing the health, safety, welfare and convenience of persons in or about buildings, and of others who may be affected by buildings or matters connected with buildings;
(b) furthering the conservation of fuel and power, and
(c) preventing waste and undue consumption, misuse or contamination of water.

The section also makes provision for, amongst other things, exemption type buildings, relaxation of the regulations, consultation with the fire authority, and general matters concerning the Building Regulations.

Part II – Supervision of Building Work etc other than by the Local Authority

Provision is made for the introduction of Approved Inspectors. It also states that a Notice (Initial Notice) should be jointly served on the local authority by a person who intends to carry out work and his/her Approved Inspector.

An Approved Inspector may be approved by

1. the Secretary of State, or
2. a Body (corporate or unincorporate) which is designated by the Secretary of State.

Note: The National House Builders Council's subsidiary company, NHBC (Building Inspection) Services Ltd, now operates as an Approved Inspector — awarded by the Secretary of State — whilst designated bodies (CIOB, RIBA, RICS, etc) have the status to approve individuals as inspectors. These bodies may also lay down limitations on approvals and designations of inspectors for certain types of work and development.

This section deals in detail with the notices and types of certificates used relating to building regulations applications. It also allows certain public bodies and nationalised industries to act as inspectors over their own work. These bodies, however, must first seek approval from the local authority in which their building work is to be executed, in the same manner as other approved inspectors.

The Building (Approved Inspectors, etc) Regulation 1985 is essential reading for exact details on Part II.

Part III – Other Provisions about Buildings

This part of the Act makes provisions for a local authority to serve notice on individuals of existing buildings to provide adequate drainage, associated conveniences, water and other important requirements to premises etc, food storage accommodation, means of escape in case of fire. Where buildings are defective and dangerous the local authority may also serve notice on individuals to have the hazard rectified. If the work is not done within the time specified, the local authority may do it themselves and recover the expense from the person in default.

Part IV – General

Amongst other provisions, the local authority is given the power to enter buildings and to execute work to make dangerous buildings safe, provided that the laid down procedure is followed.

Part V – Supplementary provisions and schedules

This part gives details of the transitional provisions, as the Act comes into force, and a schedule which covers amendments and repeals of other Acts of Parliaments due to the introduction of this 1984 Act.

Regulations Relating To Building Control

With the introduction of the Building Act 1984 the Secretary of State was empowered to make building and other regulations, such as:

- The Building Regulations 1985
- The Building (Approved Inspectors, etc) Regulations 1985
- The Building (Prescribed Fees) Regulations 1985
- The Building (Inner London) Regulations 1985
- The Building (Disabled People) Regulations 1987

 Only the Building Regulations 1985 will be briefly dealt with here.

The Building Regulations 1985 (containing Regulation No 1 to No 20, including Schedules No 1 to No 4)

These statutory instruments are divided into the following Parts and Schedules (also refer to the Building Regulations "Manual").

Part I: General

1. *Title, commencement and application*

A statement is made that the Regulations would come into effect in November 1985.

2. *Interpretation*

This regulation details the meaning of the terms used in the document. For example, a dwelling includes a dwelling house and a flat.

Part II: Control of Building Works

3. *Meaning of building work*

The erection, extention, alteration, etc of a building is classed as building work, including such operations as cavity insulation and underpinning.

4. *Requirements relating to building work*

Building work is to comply with the requirements in schedule 1 and 2.

Note: Approved Documents (13 in number) have been produced to guide individuals in their attempt to comply with schedule 1 which is divided into parts A to M (see later details).

5. *Meaning of "material change of use"*

A change occurs when a building is used as a dwelling, flat, hotel, etc where the new use is different to the previous use.

6. *Requirements relating to material change of use*

Building work should comply with the relevant requirements of schedule 1 of the regulations.

7. *Materials and workmanship*

Work should be carried out with proper materials and in a workmanlike manner. In support of this regulation an Approved Document details the way in which to control House Longhorn Beetle, High Alumina Cement, etc.

8. Limitations on requirements

Although there is an obligation to comply with regulation 1, building work need only secure reasonable health and safety conditions for users of the building and third parties.

9. Exempt buildings and works

A list of exempt buildings is given in schedule 3 of the regulations. Conservatories, porches, car ports (open at least two sides) at ground level where the floor area does not exceed $30m^2$, and a number of other buildings.

Part III: Relaxation of Requirements

10. Power to dispense with or relax requirements

The local authority has delegated responsibility from the Secretary of State to dispense with or relax the regulations where necessary.

Part IV: Notices and Plans

11. Giving of a building notice or deposit of plans

If the local authority is chosen as the authority for building control (instead of private control) by persons who intend to carry out build work or make a material change of use, a Building Notice should be submitted, or plans (Full Plans) should be deposited for the proposals to the local authority.

12. Particulars and plans where a building notice is given

A Building Notice should state the name and address of the person intending to have the building work done and should include other relevant details depending on whether the work relates to:

1. the erection/extension of a building.
2. the insertion of cavity insulation.
3. the provision of hot water storage.

Other important data is given.

13. Full Plans

Full Plans should be deposited with the local authority, in duplicate, accompanied by a statement that they are deposited in accordance with the Building Regulations 1985.

14. Notice of commencement and completion of certain stages of work

The person undertaking the building work should give the local authority in writing or other means:

(*a*) a minimum of 24 hours notice of commencement of work.
(*b*) a minimum of 24 hours notice before covering up foundation excavations, any foundations, any DPC or oversite material or concrete.
(*c*) a minimum of 24 hours notice before covering up a private sewer and for inspection of any haunching.
(*d*) a minimum of 7 days notice after covering up any drain, etc (as in c) with concrete or other material. No account should be taken of weekends and bank holidays in periods of notice.
(*e*) a minimum of 7 days notice (where a building is constructed) before occupation, and a maximum of 7 days notice after completion of the building or other work.

Part V: Miscellaneous

15. Testing of drains and private sewers

The local authority may test any drain, etc to see if it complies with the Building Regulations 1985.

16. Sampling of materials

The local authority may take samples of materials to see if they comply with the regulations.

17. Supervision of building work other than by the local authority.

Regulations 11, 14, 15 and 16 shall not apply in respect of any work specified in an Initial Notice or a Public Body Notice.

Regulation 15 and 16 shall not apply in respect of any work in relation to a Final Certificate given, or a Public Body Final Certificate given and which has been accepted by the local authority.

18, 19, 20. Repeals, Revocations and Transitional Arrangements

Special arrangements are made relating to the Building Act 1984

Schedule 1 — Requirements — regulation 4 and 6

Although the requirements of the Building Regulations 1985 are itemised within this schedule, advice and guidance on how to achieve such requirements is given in separate supporting documents which are:

- MANUAL TO THE BUILDINGS REGULATIONS 1985
- MANDATORY RULES FOR MEANS OF ESCAPE IN CASE OF FIRE - PART B1

■ APPROVED DOCUMENTS (with sub-divisions):

1 MATERIAL AND WORKMANSHIP — to support Regulation 7
2 PART A: STRUCTURE — with sub-divisions of —
A1/2: Loading and ground movement
A3: Disproportionate collapse
3 PART B: FIRE — with a sub-division of —
B2/3/3: Fire spread
4 PART C: SITE PREPARATION AND RESISTANCE TO MOISTURE — with sub-divisions of —
C1/2/3: Site preparation and contaminants
C4: Resistance to weather and ground moisture
5 PART D: TOXIC SUBSTANCES — with a sub-division of —
D1: Cavity insulation
6 PART E: SOUND — with a sub-division of —
E1/2/3: Airborne and impact sound
7 PART F: VENTILATION — with sub-divisions of —
F1: Means of ventilation
F2: Condensation
8 PART G: HYGIENE — with sub-divisions of —
G1: Food storage
G2: Bathrooms
G3: Hot water storage
G4: Sanitary conveniences
9 PART H: DRAINAGE AND WASTE DISPOSAL — with sub-divisions of —
H1: Sanitary pipework and drainage
H2: Cesspools and tanks
H3: Rainwater drainage
H4: Solid waste storage
10 PART J: HEAT PRODUCING APPLIANCES — with a subdivision of
J1/2/3: Heat producing appliances
11 PART K: STAIRWAYS, RAMPS AND GUARDS — with sub-divisions of —
K1: Stairways and ramps
K2/3: Pedestrian and vehicle barriers
12 PART L: CONSERVATION OF FUEL AND POWER — with sub-divisions of —
L2/3: Resistance to the passage of heat
L4: Heating system controls
L5: Insulation of heating services

Note: L1 is contained in the Building Regulations 1985, Schedule 1 and is the Interpretation sub-division for PART L.

13 PART M: ACCESS FOR DISABLED PEOPLE — with sub-divisions of —
M2: Means of access
M3: Sanitary conveniences
M4: Audience or spectator seating

 Schedule 2 — Facilities For Disabled People — Regulation 4

This schedule has now been superceded by PART M above due to the introduction of the Building (Disabled People) Regulations 1987.

Schedule 3 — Exempt Buildings And Work — Regulation 9

Buildings which are defined as being outside building control under the Building Regulations 1985 are given in this schedule. Examples are: certain extensions and small buildings, buildings not frequented by people, etc.

Schedule 4 — Revocations — Regulation 19

It states that the Building Regulations 1985 now supercede the Building Regulations 1976 and its amendments.

6.3 Submission of applications, approvals and inspections

Building work for which approval is required under Part II: Control of Building Work (Building Regulations 1985) is defined as:

1. the erection or extension of a building (depending on the building type and size);
2. the material alteration of a building (internal and external structural alterations, cavity insulation, underpinning);
3. the provision, extension or material alteration of a controlled service or fitting in or in connection with a building (heat producing appliances, non-vented hot water storage systems);
4. work required in the material change of use of a building (shop into house, building into a flat, etc).

Other buildings and work which are exempt from building regulation control are listed in Schedule 3 of the Building Regulation 1985, Classes I to VII., ie: temporary buildings and mobile homes, greenhouses and agricultural buildings, a building the construction of which is subject to the Explosives Acts, etc.

The control of building work etc, was mainly the responsibility of local authorities in England and Wales, but under the Building Act 1984 [see also the Building (Approved Inspectors) Regulations 1985] provision is made for the supervision of building work by other individuals or bodies within certain limitations. Supervision, therefore, may be undertaken by one of the following:

- A Local Authority
- An Approved Inspector (an individual)
- A Corporate Body (the NHBC)
- A Public Body (a Nationalised Industry or similar body)

Where building work is proposed by persons/organisations and the local authority is chosen to supervise the work, one of the following must be served the choice of which is dependant on the type of work and class of building involved:

- Building Notice

Form in Fig 6.3.1 (as laid down within the Regulations) including plans (drawings), particulars about the work, and statements given in paragraph 1 to 4 under Building Notice previously dealt with.

Rejection or passing of plans/applications should be made by the local authority within 5 weeks, or up to 2 months if the extended period is agreed because the local authority is unable to deal with a submission within the statutory time due to pressure of work.

Building operations should be commenced by an applicant within 3 years of the passing of the plans or the local authority could withdraw approval and a new submission would have to be made.

Inspections

When approval of plans etc is given by a local authority, the letter of approval would be accompanied by Inspection Notices (see Fig 6.3.2). At each recognised inspection stage of the building operations, a card would then be completed for submission to the authority. The Building Control Officer would then make himself/herself available to inspect the work at that point of the operations.

The stages at which inspection is made depends largely on the type and size of the project. More inspections will be necessary to foundations which are programmed for stage completion, unlike a single house where foundations can be excavated and concreted in a single day, thereby requiring an inspection visit by the building controls officer for that part of the work.

Notices should reach the technical services department in sufficient time to allow inspections to be carried out before the subsequent stages of the work commences or before work is covered up. The minimum periods of notice to enable the building controls officers to do their duty are given on Building Inspection Notice Fig. 6.3.2.

The first notice of commencement of work warns the technical service department that building operations are about to commence, and that they will have to be prepared for visits to the site in the near future. The responsibility for completion of the notices and for posting them to the technical services department lies with the builder.

Other officers involved in the process

When drawings are submitted to the local authority for Building Regulations approval, the following local authority officers will be available to check that the drawings conform with the Acts of Parliament most associated with their responsibilities. They are:

1. Town Planning Officer.
2. Public Health Officer.

FULL PLANS NOTICE

SEESIDE BOROUGH COUNCIL

Department of Technical Services
MILLMEAD HOUSE HIGH ST
SEESIDE, SUSSEX SE23 X7
Telephone Seeside 23456

NOTICE OF INTENTION TO ERECT, EXTEND, OR ALTER A BUILDING,

EXECUTE WORKS, INSTALL FITTINGS, OR MAKE A MATERIAL CHANGE OF USE OF AN EXISTING BUILDING

THE BUILDING REGULATIONS 1985

To: **SEESIDE** BOROUGH COUNCIL

I/We hereby give notice of intention to carry out the work set out hereunder and deposit the attached drawings an other documents in accordance with the requirements of Regulation II (1) (b).

Date .... Signed ....

Name and address of person or persons on whose behalf the work is to be carried out IN BLOCK LETTERS PLEASE ....

....

....

Tel. No. ....

If signed by Agent Name and Address of Agent IN BLOCK LETTERS PLEASE ....

....

....

Tel. No. ....

1. Address or location of proposed work ....

....

2. Description of proposed work ....

....

3. (a) Purpose for which the building or extension will be used ....

....

(b) If existing building, state present use ....

....

4. Has planning permission been sought? YES / NO
5. Means of water supply ....
6. Means of foul water disposal ....
7. Means of surface water disposal ....
8. 70% of total estimated cost of work if applicable £ ....
9. Amount of fee submitted £..
10. Do you agree to a Conditional Approval if the Council considers that it is appropriate to this particular application? YES / NO
11. Do you agree to an extension of time if this is required by the Council? YES / NO

NOTE: Two copies of this Notice should be completed and submitted together with plans in duplicate in accordanc with the requirements of Regulation 13.

Fig. 6.3.1

3. Fire Officer.
4. Highways Officer/Engineer.
5. Drainage Engineer.
6. Petroleum Officer.
7. Environmental Health Officer.

Unauthorised building

Work should not normally commence until an application has been approved. If, however, work is started and the plans are rejected, the applicant must take measures to make the building conform to the Regulations which may require alterations to work already executed. This would also apply to a building which was erected without any application having been made. If the owner refuses the request by the local building controls officer to make the building comply with the regulations, then, as a last resort, action can be taken in the form of a Court Order to have the building removed.

■ *An Approved Inspector* (an individual)

Where a person or organisation proposes to undertake building work etc, and an Approved Inspector is nominated to supervise the work, the following procedure should be adhered to under the Building (Approved Inspector, etc). Regulations 1985:

1. An Initial Notice

Should be served on the local authority jointly by the developer and approved inspector accompanied by plans (drawings) and relevant documents, and a declaration of insurance, signed by the insurer, to show it applied to the work.

The local authority has 10 days within which to reject the notice (usually made if the documents etc are not in order).

If the Initial Notice is accepted by the local authority then the Approved Inspector becomes responsible for the supervision of the building work to the exclusion of the local authority.

2. Plans Certificate

Although such plans certificate state that the drawings/plans for the building work are not defective and do not contravene The Building Regulations etc, the Approved Inspector does not have to submit the Certificate to the local authority unless the developer requests it. A Plans Certificate is then supplied to both the developer and local authority.
Note: There is a provision for the submission of a combined Initial Notice and Plans Certificate.

3. Final Certificate

Where the Improved Inspector is satisfied that the work has been completed, this certificate should be submitted to the local authority and the developer. The authority has 10 days within which to object the certificate if necessary. The Initial Notice will then cease to apply.

Side 2

Received

Inspected

Remarks

Planning Officer

Council Offices

Seeside

Sussex, SE2 3XY

Side 1

Seeside Borough Council

BUILDING INSPECTION NOTICE

Situation ____________ Plan No. ____________

The undermentioned work will be ready for inspection on

____________ 19__

[Tick appropriate section]

1. Commencement
2. Excavation for foundations
3. Foundation concrete
4. Damp proof course
5. Oversite filling
6. Drains before covering
7. Drain test after covering
8. Occupation of building
9. Completion of building

Notice 1 is required at least 48 hours before commencement.
Notices 2 and 6 are required 24 hours before inspection.
Notice 7 required not more than 7 days after covering.
Notice 8 required not less than 7 days before occupation.
Notice 9 required not more than 7 days after completion.

Name ____________

Address ____________

Nine copies are issued which are completed and returned at each work stage

Fig. 6.3.2

- *A Corporate Body* (NHBC etc)

The procedure is identical to that required by an individual Approved Inspector.

- *A Public Body* (Nationalised Industries etc)

Almost identical arrangements to that of an individual Approved Inspector.

At each notice/certificate stage the Approved Inspector, Corporate Body and Public Body should consult with the Fire Authority. Prescribed sample notices and certificates are given in the Building (Approved Inspectors etc) Regulations 1985 in Schedule 2.

6.4 Dispensation and Relaxations

It is sometimes difficult, expensive and inconvenient to adhere to the Building Regulations, and while architects/designers find it more convenient to do so, by necessity it may be realised that a measure of dispensions or relaxation is essential, bearing in mind, the possibility of unusual conditions applicable at the time, in order to complete drawings to the satisfaction of a client. This may be particularly so when designing, for example a new factory where the requirements to prevent the spread of flames but which then may affect production continuity. In such a situation an application may be made to the local authority building control department for a relaxation, provided that an application shows that other protective measures will be taken to safeguard standards of health, safety, welfare and convenience of the occupants of the building (Building Act 1984). However, the Building Regulations allow for alternative proposals, provided that the relevant requirements are complied with, whilst the Building Regulations, Schedule 1, Part B — Fire: Means of Escape is supported by the document "Mandatory Rules for the Means of Escape in Case of Fire 1985", which states that if the requirements are particularly demanding, the local authority may be asked to relax or dispense with them.

Where a local authority refuses to grant a relaxation etc, there is right of appeal to the Secretary of State.

Applications for Relaxations/Dispensations

There are 2 examples of Relaxation Application Forms. The first would be used by a local authority when making its own application for a relaxation of the building regulations to the Secretary of State. The second is used by an applicant when making a application to the local authority (see Fig 6.4.1).

Appeals

If an application to dispense with or relax a building regulation requirement is refused by the local authority, them there is a right of appeal to the Secretary of State for the Environment. If on the other hand a local authorities application for a relaxation is refused, there is no right of appeal.

BUILDING ACT 1984

The Building Regulations 1985

Relaxation of Building Regulations.

APPLICATION OTHER THAN BY A LOCAL AUTHORITY

To..(insert name of Local Authority)

I/We hereby apply under section 6 of the Public Health Act 1961, for a direction dispensing with or relaxing the requirement(s) of building regulations as specified below in connection with the proposed building or works shown on the accompanying plans (see Note 1).

PARTICULARS TO BE COMPLETED

1. State briefly the nature of proposed building or works	
2. State address of premises or location of site	
3. Has the work already been carried out? (see Note 2)	
4. State the requirement(s) of building regulations sought to be dispensed with or relaxed.	

5. State grounds for the application (see Notes 3 and 4)

(continue overleaf if necessary)

APPLICANT: Full name..(Mr./Mrs./Miss)

Address ....

....

Date.......................... (Signed)..........................

[Applicant] [Authorised to sign on behalf of applicant]
(*Strike out whichever is not applicable*)

If signed by Agent: Name of agent..........................

Profession or capacity in which acting..........................

Address of agent..........................

....

Telephone Number..........................

Fig. 6.4.1

A notice of appeal must be submitted to the local authority within one month from the date of notification of refusal. A copy of the notice would be forwarded to the Secretary of State by the local authority. Alternatively, if the local authority fails to notify a decision on a relaxation application within 2 months, an applicant can proceed with an appeal in the same way as if a refusal had been given.

Chapter 7

Safety, health and welfare on-site

7.1 Historical aspects

At the beginning of the nineteenth century, there was a growing concern regarding the working hours and conditions to which children were subjected. They formed a large proportion of the working population of Great Britain. These children started work early in their lives, as there was no compulsory schooling, and were either from a working class family background, or were orphans sent as apprentices from workhouses which were mainly supervised by Parish Councils in the large cities.

Early legislation was primarily concerned with the affairs of children and their particular problems while working for some unscrupulous employers or masters. Their working hours, cleanliness of work places, and provisions for adequate ventilation were made the subject of legislation, especially for those working in cotton mills.

The Health and Morals of Apprentices Act 1802 attempted to reduce the working hours of apprentices to twelve hours, and where prentice houses were used as accommodation for the orphans, male and female, separate dormitories were made mandatory. Robert Peel (senior) was the instigator of the 1802 Act and subsequent Acts dealing with child labour conditions. His son later continued the work in addition to pursuing other philanthropic causes.

Similar Acts continued to be enacted regarding child labour, such as those of 1819, 1825, 1829 and 1831. Magistrates were given the responsibility for the inspection of factories within their jurisdiction in the earlier Acts, but numerous problems contributed to the breakdown of the system and little improvements were made in many areas.

With the introduction of the Factories Act 1833, the first four full-time factory inspectors were appointed; each being allowed to employ a small number of assistants. They were given responsibility for the inspection of factories throughout all of England, Scotland, Wales and Ireland which can be seen as no small undertaking. However, they pursued their tasks diligently and reported on conditions and accidents met with during their tours of duty. Their reports formed the basis for sound inspection techniques which followed later in the century.

Under the Factories Act 1833 the employment of children under the age of nine was forbidden, and children under thirteen years of age were limited to eight hours a day, while those from thirteen to eighteen were restricted to sixty-nine hours in any week. Later, with the introduction of the 1844 Act, it enabled inspectors to assist operatives/employees in their claims against employers/masters when injuries were sustained by them because of the lack of safety measures to machines. Emphasis was placed on the guarding of working parts on plant and machines and the working hours of children.

By now a system of education for work children had been forced on the large employers of labour, and schools were organised – usually on a modest scale. The standards however improved through the years until eventually the Education Act 1870 made education compulsory for all working children. The Education Act 1880 was the piece of legislation which prevented employers capitalising on the cheap labour of children under the age of ten – some parents being equally responsible for allowing their children to be exploited. Compulsory education for all children up to the age of ten was then altered at the turn of the century to twelve years.

As new Acts increased the power of the Inspectorate and trade unions became stronger, they demanded better conditions for everyone. Parliament passed legislation regarding safety, health and welfare at regular intervals, each time increasing the standards for the employee. The Factories and Workshop Acts from 1847 to 1961 were substantial in numbers, and at present the Factories Act 1961 is the principal Factories Act.

Until the enactment of the 1961 Act the Regulations governing working conditions on building sites were known as the Building (Safety, Health and Welfare) Regulations 1948. As separate facilities existed for civil engineering works the 1961 Act empowered the Minister to make the Construction Regulations to control works of a building, construction and civil engineering nature. Many other Regulations have been introduced by the Minister to make provisions in such areas as the use of electricity and woodworking machines.

To give added power to the Inspectorate a new all empowering Act was passed by Parliament in 1974 called the Health and Welfare at Work, etc. Act, which not only increased the employer's liability for safety measures but additionally put the onus for safety on everyone, including the operatives. This is to ensure that each one of us becomes aware of the need to be safety conscious, and failure to be so could result, in the event of an infringement of the Act, in a heavy fine or imprisonment.

7.2 Acts of Parliament and regulations

Safety, health and welfare legislation has steadily increased the awareness of everyone of the inherent risks at most workplaces: this being particularly so on building, construction and civil engineering sites. As all Statutes within the Acts must be obeyed, employers or their representatives must familiarise themselves with the various requirements laid down in the Acts.

The employer has a duty to ensure that the workplace is safe and healthy conditions are maintained for the employees, and that others, such as the general public, should be adequately protected from dangers resulting from site operations. Some Statutes in the following Acts relate specifically, in some cases, to the building, construction and civil engineering industry:

Health and Safety at Work, etc. Act 1974.
Factories Act 1961.
Offices, Shops and Railway Premises Act 1963.

Health and Safety at Work, etc. Act 1974

This Act is the principal Act concerning safety, health and welfare of employees, and in deciding good standards the various other appropriate Acts and Regulations must be referred to by the enforcement officers, employers and their safety supervisors, site managers and operatives. Ultimately, however, the legislation prior to the 1974 Act will be replaced. The responsibility for administering the Health and Safety at Work, etc. Act lies with the now established Health and Safety Commission and Executive. As new Regulations must follow, the Minister is empowered to introduce them when conditions dictate.

Factories Act 1961

For the time being this Act will continue as a basis for reasonable standards regarding safety, health and welfare, but as new Statutory Instruments come into force under the Health and Safety at Work, etc. Act, it will eventually be superseded. At present many Regulations and Orders are in operation, having first been introduced by the Minister through the powers conferred on him by the Factories Act 1961. The main Regulations in operation at present due to this Act are:

1. The Construction Regulations.

(*a*) The Construction (Health and Welfare) Regulations 1966. Briefly these make reference to the following requirements depending on the numbers employed on the site:

i Trained person and standards of training in first aid.

ii Contents of first-aid rooms and boxes (altered by the Health and Safety (First Aid) Regulations 1981).

iii Shelter and accommodation for clothing and taking of meals.
iv Numbers of WCs and wash basins.
v Protective clothing for inclement weather.

(*b*) The Construction (Working Places) Regulations 1966, this lays down Standards regarding erections and use of scaffolding and associated equipment, i.e.:
i Use of ladders – slope, support and securing.
ii Scaffold types – suspended, trestle, etc.
iii Widths of platforms on scaffolds depending on uses.
iv Provision of handrails, toe boards and barriers.
v Frequency of inspections and records.

The following two remaining Construction Regulations were made through the powers conferred on the Minister by the Factories Act 1937:

(*c*) The Construction (Lifting Operations) Regulations 1961. The construction, maintenance and inspection of lifting appliances and associated lifting gear is controlled by these Regulations to help prevent accidents on-site, and reference is made to such things as:
i Provision of cabs for drivers of lifting appliances.
ii Suitable anchorages for lifting appliances.
iii Precautions for setting up and use of tower cranes, with safe working load notices and warning bells.
iv Competent persons to operate appliances.
v Testing and reports at recognised intervals.

(*d*) The Construction (General Provisions) Regulations 1961. These Regulations outline requirements regarding the following:
i Appointment and duty of safety officer.
ii Competent workmen to perform timbering operations to excavations, tunnels, etc.
iii Fencing and safeguarding excavations.
iv Adequate ventilation to excavations and other areas of work.
v Safety precautions and inspections to cofferdams and caissons during construction.
vi Demolition and precautions to be taken (see BS Codes of Practice 94: 1971).

The Construction Regulations were metricated by the Construction (Metrication) Regulations 1984.

2. Woodwork Machines Regulations 1974.
3. Abrasive Wheels Regulations 1970.
4. Protection of Eyes Regulations 1974 (and amendment 1975).
5. Asbestos Regulations 1969.
6. Highly Flammable Liquids and Liquefied Petroleum Gases Regulations 1972.
7. Diving Operations at Work Regulations 1981.
8. Control of Lead at Work Regulations 1980.
9. The Reporting of Injuries, Diseases and Dangerous, Occurrences Regulations 1985.

10. Safety Signs Regulations 1980.
11. Fire Certificate (Special Premises) Regulations 1976.
12. Control of substances Hazardous to Health, Regulations 1989 (COSHH).
13. The Construction (Head Protection) Regulations 1989.

Under previous Factory Acts other Regulations were enacted and should be referred to depending on the type of operations being carried out on-site, such as:

1. The Work in Compressed Air Special Regulations 1958 and 1960.
2. Electricity (Factories Act) Special Regulations 1908 and 1944.
3. Petroleum Spirit (Motor Vehicles) Regulations 1929 (as amended by Petroleum Spirit (Plastic Containers) Regulations 1982).

Offices, Shops and Railway Premises Act 1963 (metricated in 1982)

This Act was amended in 1964 and further provisions were made by the introduction of the Health and Safety (First Aid) Regulations 1981. Its main sections lay down standards of working conditions expected for office workers and are as follows:

1. Cleanliness – offices to be cleaned out at least once a week.
2. Overcrowding – minimum working space for each employee is 40 sq. ft. ($3.7m^2$).
3. Temperature – minimum of 16°C within one hour after office opens.
4. Ventilation – provisions made for adequate fresh or purified air.
5. Lighting – of natural or artificial means and should be adequate.
6. Sanitary Conveniences – to be kept clean, with separate ones for each sex. Under the Sanitary Conveniences Regulations 1964 they must be provided in the following ratios: 5 for the first 100 employees and 4 for every subsequent 100.
7. Washing facilities – basins with hot and cold running water; soap and towels provided in the same ratio as sanitary conveniences.
8. Drinking Water – to be provided.
9. Accommodation for Clothing – somewhere to hang clothes not worn in working hours.
10. Seats and Sitting Facilities – provisions made for sitting when the opportunity arises, but there must be separate seats each for sedentary workers.
11. Safety – gangways, stairs, etc. should be well maintained for safe access, and working parts of machines should be fenced or guarded.
12. First-Aid Box to be provided and a person to be responsible who is trained in first-aid treatment (further details in the Health and Safety (First Aid) Regulations 1981).
13. Fire Escapes and Certificates – adequate means of escape in case of fire to be provided and a Fire Certificate is required where more than twenty persons are employed or where more than ten persons work other than on the ground floor. Reference should also be made to the Fire Precautions Act 1971 and the Health and Safety at Work, etc. Act 1974. Exemptions from the Act are given where relatives staff the business or where not more than twenty-one hours' work is undertaken in a week.

Other relevant Acts

Contractors must be aware of additional legislation which has been introduced to cover special conditions, some examples being: the Mines and Quarries Act 1954, Explosives Act 1875 and 1923, Pipelines Act 1962 and Fire Precautions Act 1971.

In an attempt to ensure that contractors and other industrial employers observe the preservation of a healthy and satisfactory environment for members of the public, the following additional legislation was passed by Parliament: Control of Pollution Act 1974, Public Health Act 1961 and Highways Acts 1980.

The Social Security Act 1975 enables employees to claim damages in cases of industrial injuries. To ensure that adequate insurance is taken out by employers to give benefits to employees and dependants in the event of injuries being sustained at work, the Employer's Liability (Compulsory Insurance) Act was enacted in 1969 and Contractors must display an insurance certificate in a prominent position for employees to see on each of their construction sites.

7.3 Responsibility of employers and employees

Employers

Under the Health and Safety at Work, etc. Act 1974, the responsibility of the employer to employees is outlined regarding health, safety and welfare. As stated in Part 1, Section 2 of the Act, it is the duty of the employer to:

1. Provide and maintain the plant for safety and without risk to health.
2. Make arrangements for safe handling, storage and transport of goods.
3. Instruct, train and supervise persons so that the health and safety of employees is safeguarded.
4. Make workplaces safe particularly regarding access and egress.
5. Provide and maintain a safe and healthy working environment with good welfare facilities.

Employers are given the responsibility to set standards to safeguard the wellbeing of all those within his/her employment, and previously many of them have prudently prepared policies to outline clearly to the employees the firm's thoughts, particularly regarding what is expected from them on the subjects of safety, health and welfare. Part 1, Section 3 now instructs all employers to prepare a written Health and Safety Policy Statement which should include the arrangements for carrying out the policy. Modifications should be made to this document due to changing circumstances and should be brought to the notice of all employees (see Fig. 7.3.1).

It is the duty of the employer to consult with the employees' safety representatives, who may be nominated through the unions, with a view to maintaining and promoting adequate safety and health measures. This is normally undertaken by forming a safety committee within the works or on the larger sites.

Building Contractors Ltd

STATEMENT OF HEALTH AND SAFETY AT WORK POLICY

Health and safety at work etc., Act 1974

Company objective

The promotion of health and safety measures is, a mutual objective for the Company and for its employees at all levels. It is the intention that all the Company's affairs will be conducted in a manner which will not cause risk to the health and safety of its members, employees or the general public. For this purpose it is the Company policy that the responsibility for health and safety at work will be divided between all the employees and the Company in the manner outlined below.

Company's responsibilities

The Company will, as a responsible employer, make every endeavour to meet its legal obligations under the Health and Safety at Work, etc. Act to ensure the health and safety of its employees and the general public. Particular attention will be paid to the provision of the following:

1. Plant equipment and systems of work that are safe;
2. Safe arrangements for the use, handling, storage and transport of articles, materials and substances;
3. Sufficient information, instruction, training and supervision to enable all employees to contribute positively to their own safety and health at work and to avoid hazards;
4. A safe place of work, and safe access to it;
5. A healthy working environment;
6. Adequate welfare facilities.

Note: Reference should be made to the appropriate safety, etc. manuals.

Employees' responsibilities

Each employee is responsible for ensuring that the work which he/she undertakes is conducted in a manner which is safe to himself/herself, other employees and to members of the general public, and for obeying the advice and instructions on Safety and Health matter issued by his/her superior. If any employee considers that a hazard to health and safety exists it is his/her responsibility to report the matter to his/her supervisor or through his/her Union Representative or such other person as may be subsequently defined.

Management and supervisors' responsibilities

Management and supervisors at all levels are expected to set an example in safe behaviour and maintain a constant and continuing interest in employee safety, in particular by:

1. Acquiring the knowledge of health and safety regulations and codes of practice necessary to ensure the safety of subordinates' workplace;
2. Acquainting subordinates with these regulations on codes of practice and giving guidance on safety matters; and
3. Ensuring that subordinates act on instructions and advice given.

General Managers are ultimately responsible to the Company for the rectification or reporting of any safety hazard which is brought to their attention.

Joint consultations

Joint consultation on health and safety matters is important. The Company will agree with its staff or their representatives adequate arrangements for joint consultation on measures for promoting safety and health at work and make and maintain satisfactory arrangements for the participation of their employees in the development and supervision of such measures. Trade Union representatives will initially be regarded as undertaking the role of Safety Representatives envisaged in the Health and Safety at Work, etc. Act. These representatives share a responsibility with management to ensure the health and safety of their members and are responsible for drawing the attention of management to any shortcomings in the Company's health and safety arrangements. The Company will in so far as is reasonably practicable provide representatives with facilities and training in order that they may carry out this task.

Review

A review, addition or modification of this statement may be made at any time and may be supplemented as appropriate by further statements relating to the work of particular departments and in accordance with any new regulations or codes of practice.

This policy statement will be brought to the attention of all employees.

Fig. 7.3.1

In promoting a Safety Policy the employer should ensure that standards are maintained by insisting that executives/managers should be seen to be safety conscious and that more than just lip-service is being applied.

In carrying out their responsibilities employers should make booklets and guides available to supervisors, etc. such as:

1. *Construction Regulations Handbook* by RoSPA.
2. *Construction Safety*, by BEC.
3. *Guide to the Construction Regulations 1961–6*, by FCEC.
4. Technical Safety Pamphlets by RoSPA.
5. *The ABC of Offices, Shops and Railway Premises Act*, by RoSPA.
6. Safety Manuals – for the erection of tower cranes; inspection of trench timbering; erection of falsework, etc.

Further duties of employers are to display at the works the Prescribed Notices and keep up-to-date Prescribed Registers and Certificates (see part 7.4 of this chapter). Notifications are also necessary under the Factories Acts, etc. to the Factory Inspectorate, Employment Medical Adviser, Inspector of Explosives and others such as the Local Health Authority or Environmental Health Department of the Local Authority.

Appointment of Safety Supervisors

When more than twenty employees are employed a contractor has the duty under the Construction (General Provisions) Regulations 1961 to appoint a qualified Safety Supervisor to exercise observance of the Acts and Regulations governing safety, health and welfare. An alternative which in the main is made available to small firms and which is less expensive is to join a Safety Group, and by subscribing towards the cost of employing Safety Officers within the Group will be allocated the services of the Officers/Advisers as an assurance of keeping within the law, but above all, to help prevent accidents.

The few Safety Officers employed within the Safety Group act as advisers to each contractor-member of the Group. The sites of each contractor are visited in rotation and reports are prepared by the Safety Officers on conditions prevailing. If unfavourable reports are presented to contractors, it then enables them to take measures rectifying the faults to keep within the Law.

Safety Supervisors can be employed directly by firms and act as advisers operating similarly to those in Safety Groups, or they can have executive powers and thereby can insist, instead of advise, on safety measures to be taken by, for example, a site manager. Because of the importance placed on site safety executive power given to Safety Supervisors works well in some organisations but in others leads to conflict as the Site Manager becomes confused by having to take instructions from more than one superior.

Safety Supervisors employed by a contractor may also be employed in other tasks whilst a Safety Officer is employed for one task only – that regarding safety and health. Both types of officers advise the contractor on safety, etc. matters. It would be expected that Safety Supervisors/Officers would be corporate members of the Institute of Occupational Safety and Health.

Employees

All reasonable care should be exercised by employees while at their workplaces regarding their own and fellow employees' safety, under the Health and Safety at Work, etc. Act 1974. Observance of the various relevant Acts and Regulations is also the responsibility of the employee, although not to such a degree as employers.

Other documents which stress the importance of safety precautions particularly relating to taking reasonable measures by using safety helmets, protective clothing and safety equipment which are normally provided by the employer, are the various Working Rule Agreements – these documents having been compiled by the employer and employee representative bodies (unions): separate agreements exist for the building and civil engineering industries.

Where a danger exists it is reasonable to expect an employee to report the matter to enable steps to be taken to neutralise it. It is in each employee's interest to be vigilant because failure to be so could contribute to the numbers of serious accidents reported each year in the Annual Report of HM Chief Inspector of Factories.

7.4 Notices, registers and notifications

It is essential that on each construction site where either building or civil engineering work is undertaken that certain documents are made available for operatives to see, and where observance of safety, health and welfare can be recorded as laid down by the Construction Regulations and authorised under the Factories Act 1961. If a site is small, then most of the documents, particularly the Registers, would be kept at the main office and should be made available to enable the Factory Inspectorate to check that conditions regarding safety, etc. are being observed by the employer. On large sites the Registers should be kept up to date by the site manager and other persons appointed by the employer to operate safety checks, particularly the Safety Supervisor/Officer. It should be noted that under the Health and Safety at Work, etc. Act 1974 the Inspectors appointed by the Health and Safety Executive may also wish to inspect the written Policy Statement of a business.

Prescribed Forms/Notices

Where no prescribed notices or forms exist the relevant regulations should be made available at site level as described in Section 7.2 commencing with the following:

1. The Construction (General Provisions) Regulations 1961 (Statutory Instruments 1961, No. 1580).

2. The Construction (Lifting Operations) Regulations 1961 (SI 1961, No. 1581).
3. The Construction (Working Places) Regulations 1966 (SI 1966, No. 94).
4. The Construction (Health and Welfare) Regulations 1966 (SI 1966, No. 95), and the Construction (Health and Welfare) (Amendment) Regulations 1974 (SI 1974, No. 209).

The contents of these documents are outlined in Section 7.2, and other Regulations controlling working conditions of a special nature are also listed and could apply:

5. Abstract of the Factories Act 1961 for Construction Operations and Works of Engineering Construction (Form F3).
6. Abstract of the Offices, Shops and Railway Premises Act 1963 (Form OSR 9) – but only if site office is used for more than six months or an existing building is used for more than six weeks.
7. The Electricity Special Regulations 1908 and 1944 (Form 954).
8. Asbestos Regulations 1969 Notice F2358.
9. The Woodworking Machines Regulations 1974 (Form 2470).
10. Highly Flammable Liquids and Liquefied Petroleum Gases Regulations 1972 (Form 2440).
11. The Abrasive Wheels Regulations 1970 (Forms 2345, 2347 and 2351).

Once again the numbers and types of notices depend on the operations being executed.

Prescribed Registers and Certificates

During the course of construction operations records are kept of inspections and tests, and reports are made of accidents including other occurrences which may take place on site, particularly on the larger and lengthier contracts. The most important of the Registers and Certificates are listed as follows:

1. The General Register for Building Operations and Works of Engineering Construction (F36). Part 1 is a single page to show the name and address of the employer and nature of work embarked on.
2. Records of Weekly Inspections, Examinations and Special Tests (F91 (Part 1)) regarding scaffolding, excavations and hoists for carrying persons (see Fig. 7.4.1 (a), (b) and (c)).
3. Records of Reports (F91 (Part 2)) of Thorough Examinations on lifting appliances, hoists, chains and heat treatment of lifting gear (see Fig. 7.4.2 (a) and (b)).
4. Accident Book (Form B1 510) which must be available on-site to satisfy the Social Securities Act 1975.
5. Register and Certificate of Shared Welfare Arrangements (Form F 2202) – see Fig. 7.4.3.

Name or title of employer or contractor

Factories Act 1961

Address of site

Construction (Working Places) Regulations 1966

SECTION A

Work commenced—Date

SCAFFOLD INSPECTIONS

Reports of results of inspections under Regulations 22 of scaffolds, including boatswain's chairs, cages, skips and similar plant or equipment (and plant or equipment used for the purposes thereof)

Location and description of scaffold, etc. and other plant or equipment inspected	Date of inspection	Result of inspection State whether in good order	Signature (or, in case where signature is not legally required, name) of person who made the inspection
(1)	(2)	(3)	(4)

NOTES TO SECTION A

(1) *Short check list—at each inspection check that your scaffolding does not have these faults:*

		Week 1	2	3	4			Week 1	2	3	4			Week 1	2	3	4
FOOTINGS	Soft and uneven No base plates No sole boards					BRACING 'Facade and ledger'	Some missing Loose					TIES	Some missing Loose				

Fig. 7.4.1 (a)

Name or title of employer or contractor

Factories Act 1961

Construction (Lifting Operations) Regulations 1961

Address of site

LIFTING APPLIANCES

SECTION C

Aerial cableways, aerial ropeways, crabs, cranes, draglines, excavators, gin wheels, hoists, overhead runways, piling frames, pulley blocks, sheer legs, winches.

Work commenced—Date

Reports of the results of every inspection made in pursuance of Regulation 10(1)(c) or (2) of a lifting appliance or in pursuance of Regulation 30(1) or (2) of an automatic safe load indicator

Description of Lifting Appliance and Means of Identification	Date of Inspection	Result of Inspection (including all working gear and anchoring or fixing plant or gear, and where required the automatic safe load indicator and the derricking interlock) State whether in good order	Signature (or, in case where signature is not legally required, name) of person who made the inspection
(1)	(2)	(3)	(4)

Fig. 7.4.1 (b)

Name or title of employer or contractor

Factories Act 1961

Address of site

Construction (Lifting Operations) Regulations 1961

SECTION F

Work commenced—Date

HOISTS USED FOR CARRYING PERSONS

Reports of the results of every test and thorough examination made in pursuance of Regulation 46(1)(b) of a hoist used for carrying persons

Description and Means of Identification	Maximum height of travel of the cage as tested (Feet)	Date when last erected or height of travel last altered	Date of test and examination	Result of test and examination State whether in good order	Signature of person who made or who was responsible for carrying out the test and examination
(1)	(2)	(3)	(4)	(5)	(6)

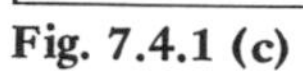

Fig. 7.4.1 (c)

4

Name or title of Employer or Contractor

Address of Registered or Head Office or Address of Site

Lifting Appliances (except Hoists)

Factories Act 1961 Section G

Reports of results of thorough examinations
(i) every fourteen months;
(ii) after substantial alteration or repair

Particulars prescribed by the Secretary of State in pursuance of regulation 28 of the Construction (Lifting Operations) Act 1961

Identification 1	Date of last previous thorough examination 2	Defects noted and alterations or repairs required to the lifting appliance and the automatic safe load indicator of a crane before the lifting appliance is put into service 3	Signature of person making examination and date of examination 4
Make and Type Maker's Number Owner's Name Owner's Number			
Make and Type Maker's Number Owner's Name Owner's Number			
Make and Type Maker's Number Owner's Name Owner's Number			

See notes on page (ii) of cover

Fig. 7.4.2 (a)

11

Name or title of Employer or Contractor

Address of Registered or Head Office or Address of Site

Hoists

Factories Act 1961 Section H

Reports of results of six-monthly thorough examinations

Particulars prescribed by the Secretary of State in pursuance of regulation 46 of the Construction (Lifting Operations) Regulations 1961

Description of hoist e.g., type, identification mark, capacity 1	Date of last previous thorough examination 2	Result of examination Enter details of repairs required or defects. If none enter "In good order" 3	Signature of person making or responsible for examination 4	Date of examination 5

See notes on page (ii) of cover

Fig. 7.4.2 (b)

6. Certificate of Tests and Thorough Examinations of Lifting Appliances:
   Hoists (Form F 75).
   Crabs, winches, etc. (Form F 80) (see Fig. 7.4.4).
   Wire ropes (Form F 87).
   Cranes (Form F 96).
   Chains and lifting gear (Form 97).
7. Register for the Purposes of the Abrasive Wheels Regulations 1970 (Form F 2346).
8. Record of Accidents, Dangerous Occurrences and Ill-Health Enquiries (Form 2509).

Notifications to HM Factory Inspectorate and others

When certain events take place on-site, it is the duty of the employer or representative to send details to the appropriate body under the various safety, health and welfare Acts and Regulations; these are as follows:

1. Notice of Building Operations or Works of Engineering Construction (Form F 10) – only sent if operations will last for more than six weeks (see Fig. 7.4.5).
2. Notice of Employment of persons in offices which will be used for more than six months and where more than twenty-one man-hours are worked weekly (Form O SR 1).
3. Notice of Accidents or Dangerous Occurrences (Form 2508) – but only for accidents leading to death or serious injury (absence from work of more than three days) and for all dangerous occurrences (see Fig. 7.4.6).
4. Notice of Taking into Employment or Transference of a Young Person (Form F 2404) – sent to the local Careers Officer.
5. Notice in Case of Poisoning or Disease (Form F 41) – one copy sent to the Factory Inspector and another to the Employment Medical Adviser (see Fig. 7.4.7).

7.5 Health and safety inspectors

In Fig. 1.4.5 one can see the extent of accidents in the construction industry. On average over the past ten years one person was killed each working day and there were 36 000 reported accidents per year requiring each casualty to spend more than three days away from work.

In addition to the stress caused to families when accidents occur, valuable working hours are lost while the injured are tended by their colleagues until such time as they are removed to the hospital for medical treatment, or until the incapacitated finally returns to work on recovery. The costs of accidents are borne by everyone – the contractor/employer, client, Government; not to mention the loss of earnings – flat rate of pay, bonus and overtime – by the operatives and their dependants.

Factories Act 1961

F 2202

CONSTRUCTION (HEALTH AND WELFARE) REGULATIONS 1966

Certificate of shared welfare arrangements made under regulation 4

Form approved by H.M. Chief Inspector of Factories

PART A

(To be completed by the contractor providing facilities and handed to the contractor for whom facilities are provided)

Name or title of Employer or Contractor providing the facilities

____

Address of Site

____

____

Name of Employer or Contractor for whom facilities are provided

	Facilities provided						
	First-aid boxes (Regs. 5(1), 10(2) and Schedule.)	Trained first-aider (Regs. 5(2), 7 and 10(2).)	Ambulance arrangements (Regs. 8 and 10(2).)	First-aid room (Regs. 9 and 10(2).)	Shelters and accommodation for clothing and taking meals. (Reg. 11.)	Washing facilities (Reg. 12.)	Sanitary conveniences (Reg. 13.)
Whether facilities provided (Yes/No)							
Date arrangements began							
Date arrangements ended							

Note

This certificate must be kept by the contractor for whom facilities are provided. It must be kept on the site or at his office and must be available for inspection by H.M. Inspectors of Factories or any employee who is affected by the arrangements.

(Signed)_______________

For and on behalf of (Name of Employer or Contractor providing facilities) _______________

Date_______________

Fig. 7.4.3

Certificate No. ....

HEALTH AND SAFETY EXECUTIVE
Factories Act 1961

THE CONSTRUCTION (LIFTING OPERATIONS) REGULATIONS 1961

Certificate of Tests and Thorough Examination of

(a) Crabs and Winches

(b) Pulley Blocks and Gin Wheels used for a load of one ton or more

(Prescribed by the Secretary of State in pursuance of Regulation 28 (5))

Owner's Name and Address:

Description of appliance(s), type and distinguishing mark	Test load applied (tons)	Safe working load (tons)	Defects noted, alterations or repairs required. (If none, enter "None")

I hereby certify that on .... the appliances described in this Certificate were tested and thoroughly examined and that the above particulars are correct.

Signature

Qualification

Person or Firm by whom the person conducting the test and examination is employed

Date of Certificate

NOTES—

A crab or winch must **NOT** be used unless a certificate of test and thorough examination by a competent person has been obtained in this form **within the previous four years and since any substantial alterations or repair affecting its strength or stability.**

A pully block or gin wheel must **NOT** be used for a load of one tonne or more unless a certificate of test and thorough examination by a competent person has been obtained in this form **previously and since any substantial alteration or repair.**

Fig. 7.4.4

F 10

Reprinted December 1979

For official use
Registered................................
Visited....................................

FACTORIES ACT 1961

Notice of building operations or works of engineering construction*

1	Name of person, firm, or company undertaking the operations or works.	
2	State whether main contractor or sub-contractor.	
3	Trade of the person, firm or company undertaking the operations or works.	
4	Address of registered office (in case of company) or of principal place of business (in other cases).	
5	Address to which communications should be sent (if different from above).	
6	Place where the operations or works are carried on.	
7	Name of Local Government District Council (in Scotland, County Council or Burgh Town Council) within whose district the operations or works are situated.	
8	Telephone No. (if any) of the site.	
9	How many workers are you likely to employ on the site?	
10	Approximate date of commencement.	
11	Probable duration of work.	
12	Is mechanical power being, or to be, used? If so, what is its nature (e.g. electric, steam, gas or oil)?	
13	Nature of operations or works carried on:	

(a) Building operations *(tick items which apply)*

Construction ....
Maintenance ....
Demolition ....

of

Industrial building ....
Commercial or public building ....
Dwellings over 3 storeys ....
Dwellings of 3 storeys or less ....
Others ....

(b) Works of engineering construction *(specify type)*

I hereby give notice that I am undertaking the building operations or works of engineering construction specified above.

Signature   Date

NOTE

* Any person undertaking any building operations or works of engineering construction to which the Act applies is required by the Act, not later than seven days after the beginning of any such operations or works, to serve on the Inspector for the district a written notice giving particulars specified in section 127(6) unless (a) they are operations or works which the person undertaking them has reasonable grounds for believing will be completed in less than six weeks, or (b) notice has already been given to the Inspector in respect of building operations or works of engineering construction already in progress at the same place. This form should be filled up and sent to HM Inspector of Factories for the district in which the operations or works are carried on.

572 8033375 200M 12/79 HGW 752

Fig. 7.4.5

Health and Safety Executive
Health and Safety at Work etc Act, 1974
Notification of Accidents and Dangerous Occurrences Regulations, 1980

Report of an accident and/or dangerous occurrence and injuries sustained

Please read the notes on pages 1 and 2 before completing this form.

Part I Administrative

1 Person or organisation reporting the accident/ dangerous occurrence

Name ______

Address ______

______

______ Postcode ______

Nature of undertaking ______

Signature of person making this report ______

Date ______

Name *(block capitals)* ______

Position in organisation (where applicable) ______

Tel no ______ Ext ______

2 Place of accident/dangerous occurrence if different from 1 (for construction sites give name of main contractor)

Name ______

Address ______

______

______ Postcode ______

Tel no ______ Ext ______

Name of site manager or other person in charge *(block capitals)*

Address *(if not as above)* ______

______

______ Postcode ______

Tel no ______ Ext ______

Part II General report of the incident

1 Date ______ 2 Time ______ *am/pm

3 Precise place, e.g. South Warehouse, No 2 Machine Shop, canteen kitchen

______

4 Was there a dangerous occurrence as defined in the regulations? yes ☐ no ☐

If yes, state type ______

5 Number of (a) deaths ______ (b) major injuries ______

6 Give a full account of the accident/dangerous occurrence, explaining so far as possible how it happened and how those killed or hurt received their injuries. Give name and type of any plant, equipment, machinery or vehicle involved and note whether it was in motion.

F2508 **delete as appropriate* Part III – overleaf

Fig. 7.4.6

Health & Safety Executive

HM Factory Inspectorate

Form prescribed by the Secretary of State

F41

Notice of case of poisoning or disease

occurring in a factory or in other premises or places to which the provisions of section 82(3) and (4) of the Factories Act 1961 apply

This box to be filled in by District Inspector

District

Number of case

1 Occupier of factory
(or person carrying on processes at docks and certain other places)

a Name

b Address

c Industry

2 Actual employer (if other than above)

a Name

b Address

3 **Place where person affected had been working**

a Address (if different from 1b above)

b Exact location

c Nature of work carried on there

4 **Person affected**

a Full name (surname first)

b Sex Age Precise occupation (avoid the term *labourer* where possible)

c Address

5 **Nature of poisoning or disease**

6 **Notifications**

a Has the case been reported to the Employment Medical Adviser?

b Has the case been reported in the General Register?

Signature of occupier, etc

Date

Notes

1 The provisions of section 82 apply not only to factories but also to certain electrical stations, the sites of building operations or works of engineering construction, railway running sheds, docks, wharves, quays and warehouses and certain work on ships in harbour or wet dock, ie constructing, reconstructing, repairing, refitting, painting, finishing or breaking up a ship or scaling, scurfing or cleaning boilers (including combustion chambers and smoke boxes) in a ship, or in cleaning oil-fuel tanks or bilges in a ship or in cleaning in a ship any tank last used for oil of any description carried as cargo, and the loading, unloading or coaling of a ship in a dock, harbour, or canal.

2 A notice in this form should be given forthwith by the occupier of the factory or premises if there occurs a case of beryllium, cadmium, lead, phosphorus, manganese, arsenical, mercurial, carbon bisulphide, aniline, or chronic benzene poisoning or of toxic jaundice or toxic anaemia, compressed air illness, anthrax or epitheliomatous or chrome ulceration.

3 Notification is also required in respect of lead poisoning occurring in any other place where persons are employed in connection with the painting of buildings or in certain processes connected with the manufacture of lead or the use of lead compounds.

4 In the case of building operations, works of engineering construction, docks, work involving lead, etc the notice should be given by the employer of the person affected.

5 The notice should be sent forthwith both to the District Inspector of Factories and also to the Employment Medical Adviser for the area in which the factory is situated. Notification to the Employment Medical Adviser is not, however, required in cases of lead poisoning where persons are employed in certain processes connected with lead manufacture or involving the use of lead compounds in places other than those specified in Note 1 (above)

Printed in England
by Robendene Ltd., Amersham
and published by
Her Majesty's Stationery Office

20p per copy
£1.50 for 25 copies (exclusive of tax)

Dd 022911 K29 12/80
ISBN 0 11 881221 1

Fig. 7.4.7

Accidents are invariably caused by carelessness. This is due to the lack of care by individuals, poor safety checks, inadequate supervision and maintenance of plant. Until recently, too little emphasis was placed by firms and employees alike on safety.

To ensure that operations are carried out, safety enforcement officers are appointed under the Health and Safety at Work, etc. Act 1974, and are entrusted to ensure that everyone is complying with the letter of the Law regarding safety, health and welfare. Workers, visitors to site and the general public require protection, and it is the main duty of each employer to look to his/her responsibility in this field.

With the formation of the Health and Safety Commission, representatives of which come from both sides of industry, the Health and Safety Executive was then created to act on behalf of the Commission. Under its umbrella the Executive controls the various health and safety inspectorates, such as:

1. Factory Inspectorate.
2. Mines and Quarries Inspectorate.
3. Explosives Inspectorate.
4. Nuclear Installations Inspectorate.
5. Alkali Works Inspectorate.
6. Pipe-lines Inspectorate.
7. Employment Medical Inspectorate, etc.

Factory Inspectors

Although these Inspectors play a prominent part in carrying into effect the provisions in the Health and Safety at Work, etc. Act 1974, Inspectors from other sections controlled by the Health and Safety Executive could, from time to time, be called to give advice on their speciality.

The Inspectors have powers conferred on them regarding safety, health and welfare, which are:

1. Power to enter premises to carry out their duties under the Act.
2. Power to take samples of anything for examination or investigation.
3. Power to require any person (employer, employee) to give information when necessary.

Other authorised persons with a specialist knowledge may be taken by the Inspectors to take photographs, measurements and recordings which may be used in investigations. A duty is imposed on an Inspector to inform the employees and employer of any matter which may be prejudicial to safety, health or welfare of the occupants of the works.

Improvement Notices

This Notice is served by an Inspector on an employer where there is a contravention of the Act and it usually stipulates a period in which the infringement should be rectified.

Prohibition Notices

Where an activity is risky to employees or others which could, in the eyes of the Inspector, lead to injury, this Notice could be served on an employer – to take effect immediately, and the activity should cease forthwith until the risk is eliminated to the Inspector's satisfaction. This could apply to badly erected scaffolding or trenches where the timbering is inadequate.

In addition to Notices being served on an individual, the person who is found to be contravening the Statutory provisions can be prosecuted and if summarily convicted can be imprisoned for a maximum of two years.

7.6 Committees and safety training

It is a recommendation that safety committees should be set up within businesses comprising representatives from the management and employees sides: this is however normally only practicable in medium to large companies. The unfortunate problem in the construction industry is that firms' activities are spread over wide areas and it may be found necessary to have works safety committees not only at head office but on each of the larger sites: the smaller sites being controlled by the head office safety committees, perhaps with one or two representatives from the sites.

Site safety committees should be allowed to meet regularly, say on a weekly/fortnightly basis, to help implement the Company Safety Policy and to make recommendations when necessary to improve it and to ensure adequate safety measures are taken on site. This will enable the employer to show that the Policy is 'seen to be done' (see The Safety Representatives and Safety Committees Regulations 1977).

Adequate provisions should be made for all employees to communicate with the committees, especially if they can highlight safety problems or risks they have found or can foresee regarding the various operations on-site, or indeed, at the head office.

Trade union representatives (site or shop stewards) are thought more suitable to represent the workforce, and members of the unions can approach their leaders with suggestions for improvements, or if necessary, suggestion boxes can be made available in canteens or other suitable positions.

Where site safety committees are formed, to get the best results, someone in authority with power to act on recommendations put forward should be nominated as chairman: the site manager or even the company's safety officer would be best suited here. Other members of the site safety committee would be the relevant foremen and representatives from each trade or union.

Awareness of the various safety Codes and company safety Manuals is essential by each member of the Committee and the priority on formation of the Committee is for the safety officer to give instructions on these reference documents.

Under the Health and Safety at Work, etc. Act 1974, the employer is made responsible to ensure that adequate instruction is given to employees to make them safety conscious. Numerous firms have operated suitable safety training schemes for many years and have set standards which the less enlightened would be well advised to observe. It is necessary, and has always been found so, to employ a trained person to act as the Safety Officer: this can be done by employing the person directly, or by employing him/her indirectly through a Group Safety Scheme. This will then ensure that the firm will get results as the appointed Safety Officer will put into motion the safety policy of the firm. The Safety Officer should be entrusted solely with matters relating to safety, health and welfare and should not be burdened with extraneous duties if he/she is to be successful. He or she should also, if possible, be a member of the Institution of Industrial Safety Officers.

The Safety Officer visits offices, workshops and siteworks and is an inspector and adviser. After each visit written reports are made out so that management can monitor safety observance by, particularly, the site managers. The other duties of the Officer are as follows:

1. To send intending scaffolders/falseworkers, etc. on recognised training courses which are either held within the company, or more usually at Training Centres run by the Construction Industry Training Board.
2. To ensure that New Starters (apprentices and trainees) receive suitable instructions on the various hazards either on-site or in the workshop by the following means:
   (*a*) Lectures.
   (*b*) Films.
   (*c*) Slides.
   (*d*) Visits – to safety exhibitions run by employers' organisations, manufacturers of safety equipment, etc. and RoSPA.
3. To arrange films and lectures for operatives on-site, say, after lunch – which may mean extending the period by fifteen to thirty minutes.
4. To display suitable posters of likely hazards or accident victims – changing them regularly which then emphasises the concern the employer shows to safety aspects.
5. To display protective clothing in the canteen, usually giving special purchase discounts to encourage use by the operatives.
6. To display safety equipment with instructions of how it is used.
7. To up-date Safety Manuals in the light of changing conditions and Legislation.
8. To show Statistic Charts illustrating the worst and most recurring types of accidents.

Chapter 8

Measurement procedure

8.1 The people who undertake measurement

The quantity surveyor

On projects of any size the architect will normally advise his client on the appointment of a quantity surveyor. Private quantity surveyors work in practices similarly to architects and are often called the professional quantity surveyor. They are paid a fee based on the total contract sum. Quantity surveyors are normally members of the Royal Institution of Chartered Surveyors (RICS). They will also employ staff at technician and student level to assist with the work.

Public undertakings and local government departments have their own quantity surveying staff to control and advise on the cost of building projects.

Measurement work undertaken by quantity surveyors is varied and interesting and will include some of the following:

Cost appraisal

There are several techniques available which enable an approximate estimate of cost to be made of building projects at sketch design stage. This can assist the architect and client in establishing the most economic design for the required building work.

A bill of quantities is measured from the drawings and sets out to measure in detail the proposed building work in a standard manner which can subsequently be priced by contractors. The priced bill is then used throughout the contract for valuation and cost control purposes. Students are strongly advised to obtain a copy for reference purposes. Building contractors or quantity surveyors will often help out with a used one if asked.

Measurement on site

Quantities are often measured provisionally and are then subject to remeasurement on site as the actual work is undertaken. This is particularly the case with sub-structure, drainage and external works.

Valuation for stage payments

Most building contracts provide for the contractor to be paid on account as the work progresses. This is usually done on a monthly basis. The quantity surveyor will normally visit the site and evaluate the work undertaken in the current month. The valuation is then prepared in a suitable form for the architect who issues a certificate authorising payment to the contractor.

Valuation of variations to the contract

Building contracts are usually subject to extensive variations. These are caused by a variety of reasons. The effect is to alter the design or specification and therefore the estimated price which the contractor originally submitted. The quantity surveyor should use his best endeavours to value the variations fairly to the satisfaction of the client and the contractor.

Fluctuations in cost of materials and labour

Because of the difficulty in predicting the rate of increase in the cost of materials and labour during a building project the common practice is for contractors to base their estimate at current prices. The contractor is then reimbursed as materials rise in cost or nationally agreed wage increases come into force. The contractor, therefore, does not have to predict the level of future inflation at the estimating stage.

The quantity surveyor must check and agree any claims for increased costs presented by the contractor.

Note: Some types of contracts do not allow for these fluctuations, but are usually limited to contracts not exceeding twelve months' duration.

Preparation of final account

At the end of a building project the quantity surveyor prepares the final account which is similar to a balance sheet, showing how the total contract sum has been calculated. Because of the effect of variations and fluctuations the final cost will normally exceed the estimated price and the client should be made aware of this fact.

The work of a quantity surveyor then can be likened to that of an accountant with an extensive practical knowledge of building. Remember also that the quantity surveyor is appointed to see fair play to both client and contractor.

The contractor's staff

To enable contractors to work profitably it is desirable that as many staff as possible should have a knowledge of measurement and contractual procedures. This will assist the contractor in making claims for various payments to which he is entitled under the terms of the contract.

Many companies will employ staff at a senior level who are members of the Chartered Institute of Building. Others may hold the national certificate or technician education certificate or have come through the trade with City and Guilds qualifications.

The contractor's surveyor: Often known as the quantity surveyor in construction companies, but to avoid confusion we will use the term 'contractor's surveyor' to distinguish this title from the previously mentioned quantity surveyor. The contractor's surveyor is employed directly by his company and his duty is entirely to his organisation. Several of his functions overlap or assist those of the quantity surveyor, these include:

1. Measurement on site.
2. Assistance with preparation of valuations with the quantity surveyor.
3. Agreeing value of variations with the quantity surveyor.
4. Preparing information with regard to fluctuation claims.
5. Agreeing the final account with the quantity surveyor.

In addition there are tasks which are completely independent of the quantity surveyor.

Cost reconciliation

This is to compare estimated costs with actual costs as the work progresses. Information can be fed back for use in estimating on new projects and also to advise management should a contract be going astray financially. For this type of information to be useful it must be available as quickly as possible.

Contract administration

Dealing with claims by nominated subcontractors and suppliers and labour-only subcontractors. Measuring and checking on site for calculation of bonus payments.

Other staff

Other members of the contractor's staff are also engaged in the measurement processes.

Estimator: The estimator must fully understand the code of measurement processes.

Buyer: The buying department often produces schedules of material requirements as well as negotiating the best possible prices for supply of goods.

 Planners: Larger companies employ planners to identify likely building programme times. By using bulk quantities of work suitable time scales can be built up.

Contract managers: Normally have overall charge of contracts and therefore need to be well informed of financial state of contracts and of the cost control techniques employed by the company.

Site staff: Site staff should not be overlooked. They should not be overloaded with paperwork, but at the same time the correct procedures with such items as daywork sheets and material transfers are an important part of the measurement process.

8.2 The preparation of a Bill of Quantities — dimensions and descriptions

The preparation and subsequent use of Bills of Quantities on construction projects is the major process involving measurement techniques.

Some uses of the bill

1. To measure the work contained in the construction or alteration of building works in a systematic and standard manner in order to obtain competitive tenders from contractors for the project.
2. To assist in preparation of valuations of work in progress in order to make stage payments to the contractor.
3. To assist in the valuation of variations occurring during the progress of the project.
4. To assist the contractor in planning resources such as materials, labour and plant.
5. To assist in the preparation of the final account at the completion of the project.
6. To assist the quantity surveyor in preparing approximate estimates for future projects.

Production of the bill

The bill is prepared from the architect's drawings by the quantity surveyor in a very exacting manner. It is prepared in accordance with a code of measurement known as **The Standard Method of Measurement of Building Works (SMM).**

Pricing the bill

Those contractors who are to compete for the work receive the Bill of Quantities with other contract documents direct from the architect.

The contractor's estimator will read through all the contract documents to get a thorough feel for the proposed work. A study of the drawings will be made. The site will be visited and a note made of any particular difficulties that may be apparent. For this purpose the estimator will often use a check list. Discussions with other members of the building team may be held to decide on the best construction methods to use. Meanwhile building material suppliers and subcontractors' quotations will be obtained for use in pricing the bill.

Having assimilated all this information the estimator can then work through the bill calculating prices for each detailed item.

Finally the bill may be totalled and a decision made as to the percentage to be added to allow for the contractor's overheads and profit. This total sum becomes the contractor's offer of a contract price which is known as the **tender price.**

Completed tenders

Tender prices usually have to be received at the architect's office on or before a specified time and date. Tenders are then opened and scrutinised. It is the usual practice to accept the lowest tender, although the client is not bound to do so. The successful contractor's priced bills are then called for and are checked to ensure there are no arithmetical errors which may make the tender uncompetitive. Assuming that there are no such errors the client and contractor may proceed to signing the actual contract to carry out the works.

Advantages of a Bill of Quantities

The student will have realised that each contractor prices an identical bill. Any variance in tender prices will be due entirely to:

1. The contractor's desire to obtain the work.
2. The contractor's ability to submit a competitive estimate.

As the quantity surveyor has prepared the bill the contractor is relieved of the function of measuring quantities and can concentrate on the build-up of the estimate alone. Had each contractor to measure their own quantities there would be much duplication of effort and there is always the possibility of error and misinterpretation. The contractor is aware that any mistakes in the Bill of Quantities can be rectified by means of a variation and therefore the possibility of errors is reduced.

Because of the expense and time involved in the preparation of Bills of Quantities they are not normally used for smaller works.

Measuring quantities

The measurement processes involved in preparing a Bill of Quantities can be divided into four main headings namely:

1. Taking-off dimensions.
2. Squaring dimensions.
3. Abstracting.
4. Writing the Bill of Quantities.

Let us examine each of these stages in a little more detail.

This is the process of measuring quantities from detailed construction drawings. The measurements are recorded on paper known as **dimension paper.** A blank page of dimension paper is shown in Fig. 8.2.1. For information purposes the use to which each column is put is also shown in Fig. 8.2.1, although these column headings would not appear in practice. Dimension paper is divided into two equal halves and is used by writing down the left-hand side of the page from top to bottom and then similarly for the right-hand side of the page. A small binding column is often provided on the extreme left-hand edge of the paper for attaching together sheets of completed dimensions. The following points should also be noted:

1. Always use the paper the correct way as Fig. 8.2.1.
2. Number each page at the foot of the description columns for reference purposes.
3. At the head of each page write the title of contract or example being worked on.

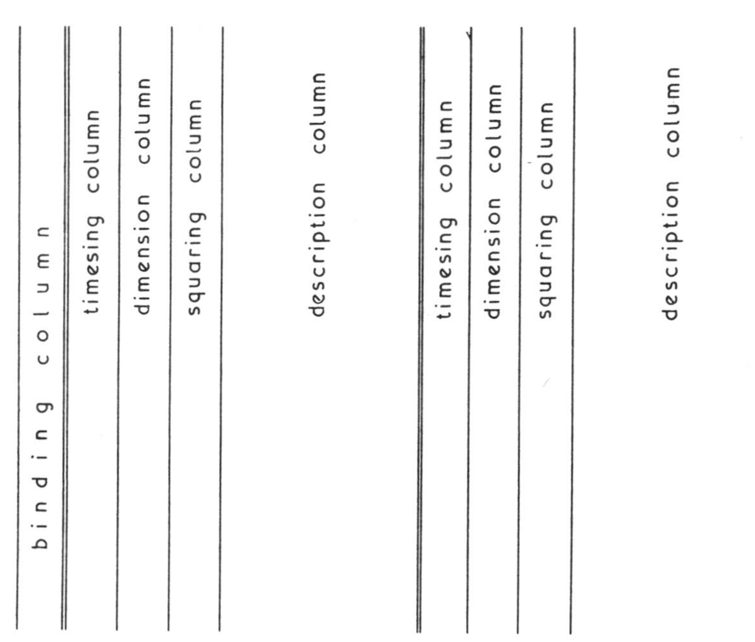

Fig. 8.2.1 Blank sheet of dimension paper

Squaring dimensions

The squaring up of dimensions is also carried out on dimension paper. Squaring is done when all dimensions for a section of the work have been completed, or on smaller works, when the whole take-off is finished. As this stage involves the multiplication or adding together of figures measured initially as dimensions, an electronic calculator will be found useful.

Accuracy is important in squaring up, as indeed it is in all stages of measurement. In a quantity surveyor's office the squaring process is often undertaken by someone other than the person who actually carried out the taking-off and is carefully checked.

Examples of squaring up are given later in this chapter.

Abstracting

The preparation of the abstract follows the squaring process. The squared dimensions are transferred from dimension paper on to abstract paper. Abstract paper is double-width A3 size. The purpose of the abstract is to collect together and to arrive at total quantities for each description of work, for the subsequent inclusion into the Bill of Quantities.

In small examples such as contained in this book, the abstracting stage can be dispensed with. Billing can be carried out direct from the squared dimensions.

Writing the Bill of Quantities

Finally the Bill of Quantities is written up in a similar form to that shown at the end of Chapter 8.

Use of dimension paper

Before actually measuring any work we must become familiar with the types of dimensions, the ways in which they are written down on dimension paper and also the other terms used in the taking-off process.

These may be listed as follows:

1. Types of dimensions
   (*a*) linear metres,
   (*b*) square metres,
   (*c*) cubic metres,
   (*d*) enumerated items.
2. The description.
3. Bracketing.
4. Squaring.
5. Timesing.
6. Dotting-on.
7. The ampersand.
8. Correction of dimensions.
9. Waste calculations.

Examples of each of these terms are now given. The descriptions are phrased from the standard method of measurement of building work. In these examples the right-hand half of the dimension paper is deleted in order that explanations may be given.

The dimension, description and bracketing

This simple example sets out to explain the use of these three terms. Assume that the sewer as detailed in Fig. 8.2.2 is to be measured between the manholes.

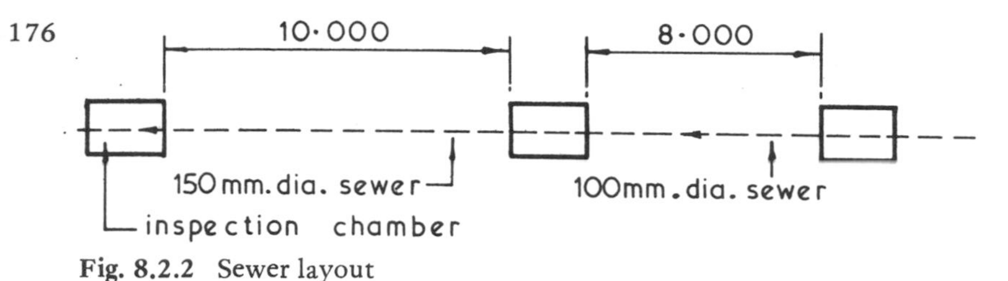

Fig. 8.2.2 Sewer layout

Stage 1. Calculate dimensions to two places of decimals and insert in dimension column. Drain and sewer runs are measured in linear metres.

Stage 2. Write suitable description of the work based on the standard method of measurement of building works.

Stage 3. Bracket dimensions and descriptions together by drawing a bracket just wide of the line separating squaring and description column.

Stage 4. Commence next dimension and description leaving space of about 40 mm.

In Fig. 8.2.3 the sewer is measured in linear metres. There is no need to state that the dimension is a linear one, as a single dimension with a line drawn under it indicates that the work has been measured linearly.

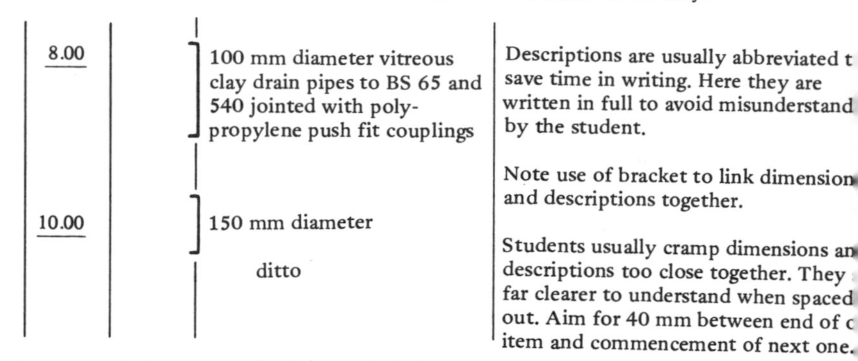

Fig. 8.2.3 Dimensions for Figure 8.2.2

Similarly, a square dimension is measured length × breadth or depth and would appear as follows:

10.00
3.00

Finally a cubic dimension is expressed as length × breadth × depth for example:

12.00
1.00
1.50

Examples of linear, square and cubic dimensions are given in Figs. 8.2.4, 8.2.7 and 8.2.10.

Other terms and methods used in the taking-off process are shown in Figs. 8.2.13 and 8.2.14.

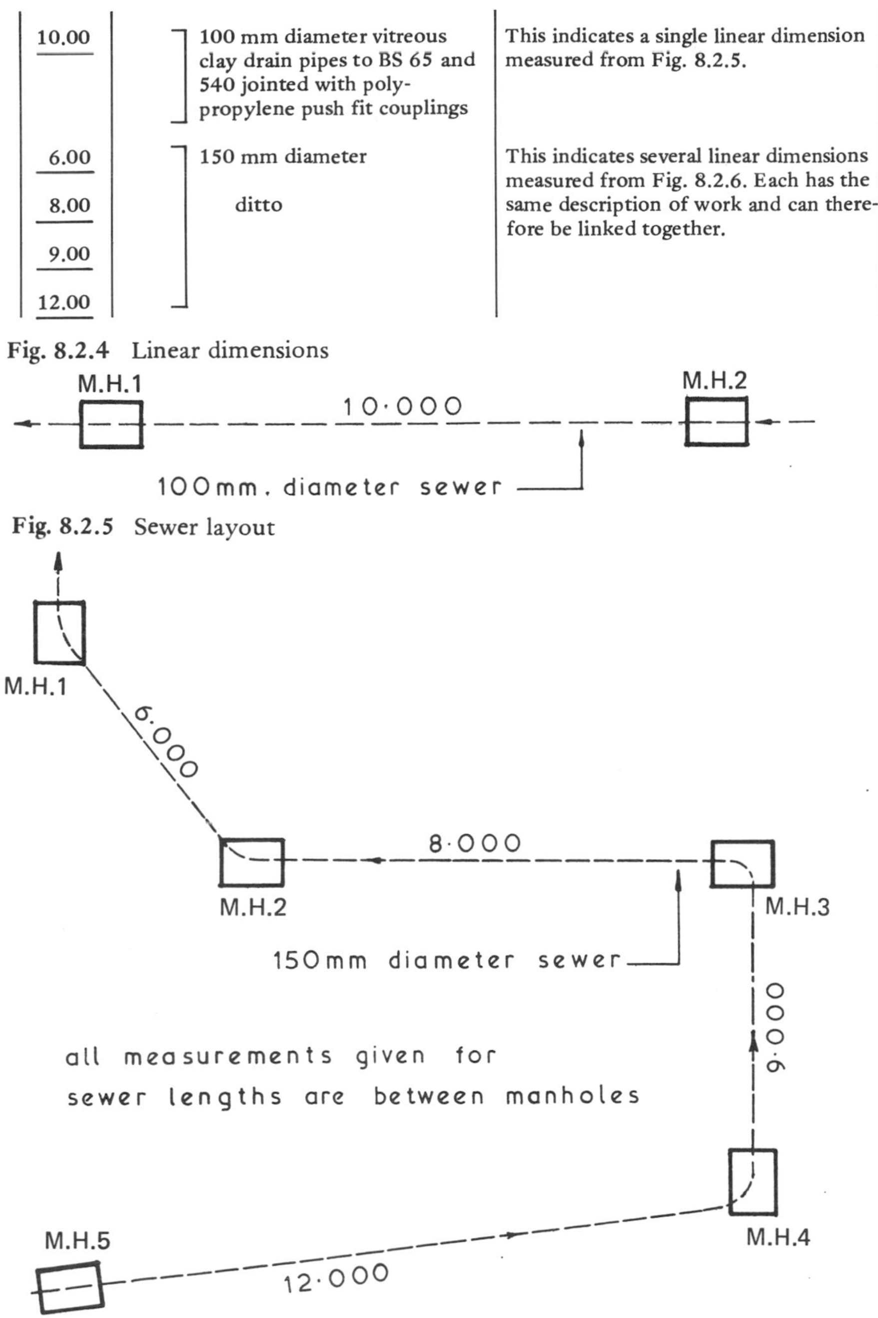

10.00	100 mm diameter vitreous clay drain pipes to BS 65 and 540 jointed with polypropylene push fit couplings	This indicates a single linear dimension measured from Fig. 8.2.5.
6.00	150 mm diameter	This indicates several linear dimensions measured from Fig. 8.2.6. Each has the same description of work and can therefore be linked together.
8.00	ditto	
9.00		
12.00		

Fig. 8.2.4 Linear dimensions

Fig. 8.2.5 Sewer layout

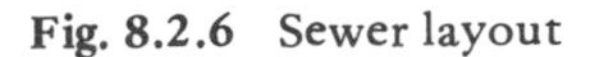

Fig. 8.2.6 Sewer layout

	Dimensions		Description	Notes
	4.00 5.00		Thermoplastic floor tiles (PC. £6 per m^2) size 300 × 300 × 3 mm bedded in approved adhesive on level screeded bed (measured separately)	This indicates a single square dimensi measured from Fig. 8.2.8. Square dimensions are expressed as length × breadth or length × depth
	5.00 3.00 4.50 4.00 4.00 3.50 4.50 3.00		Thermoplastic floor tiles as before described	This indicates several square dimensic measured from Fig. 8.2.9. Each has t same description and can therefore b linked together. Note also how the ta off can save time with the writing of descriptions where the work is of a similar nature to that previously mea Work usually measured in square met includes: 1. Brickwork and blockwork. 2. Plastering to walls and ceilings. 3. Painting and decorating to walls a ceilings. 4. Floor finishings and coverings.

Fig. 8.2.7 Square dimensions

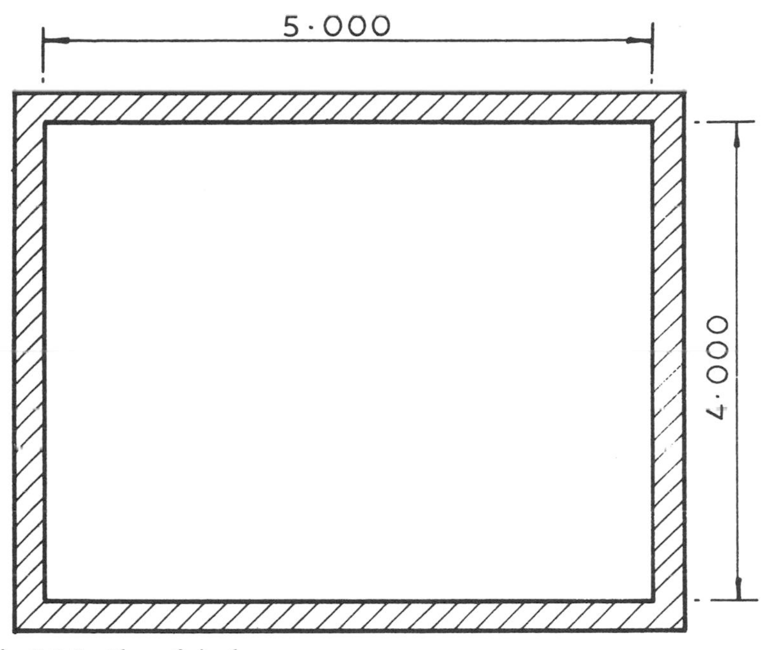

Fig. 8.2.8 Plan of single room

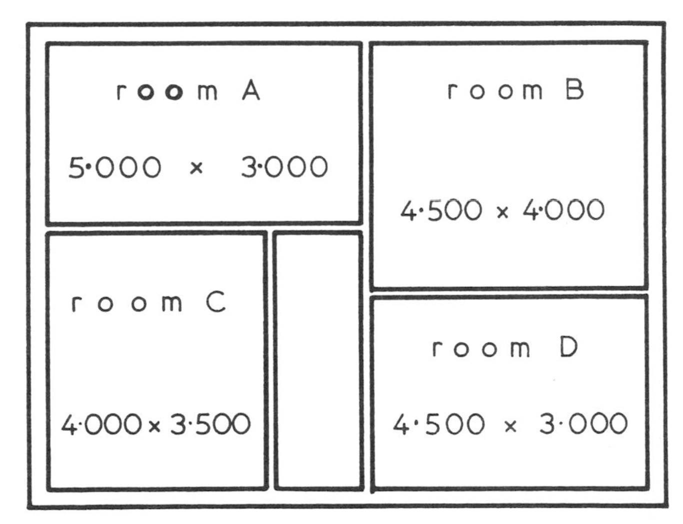

Fig. 8.2.9 Plan of building

Dimensions	Description	Notes
1.25 1.25 1.10	Excavate pit for bases of piers and the like commencing at reduced level maximum depth not exceeding 2.00 m (in 1 number)	This indicates a single cubic dimensi measured from Fig. 8.2.11. Cubic dimensions are expressed as: length × breadth × depth
1.25 1.25 1.40 1.40 1.40 1.50 1.60 1.60 1.75	Excavate pits for bases of piers as before described (in 3 number)	This indicates several cubic dimensic measured from Fig. 8.2.11. Each ha same description and can therefore linked together. Work usually measured in cubic met includes: 1. Excavation work. 2. Disposal of excavated material. 3. Filling material over 250 mm thi 4. Concrete work.

Fig. 8.2.10 Cubic dimensions

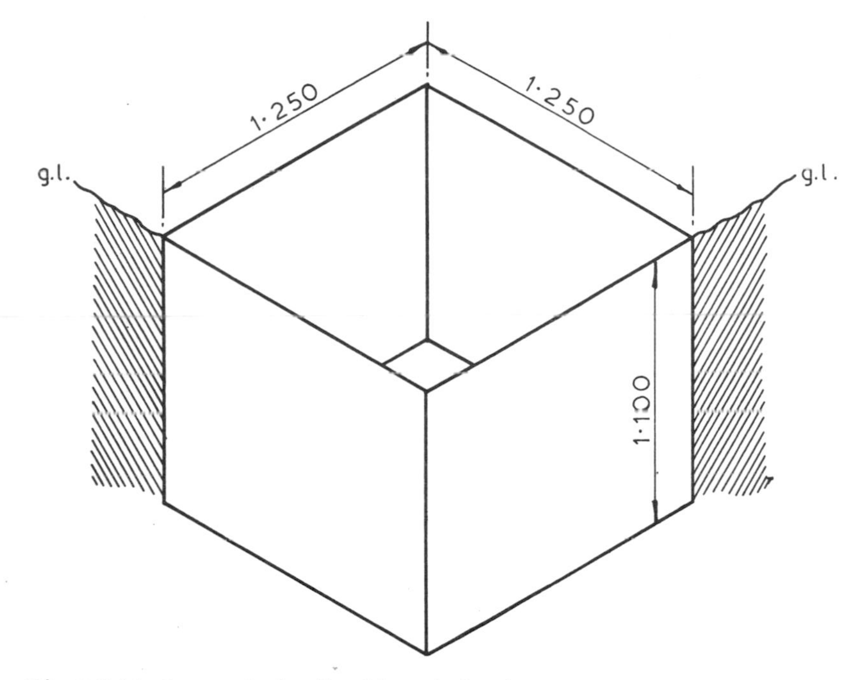

Fig. 8.2.11 Isometric details of foundation base

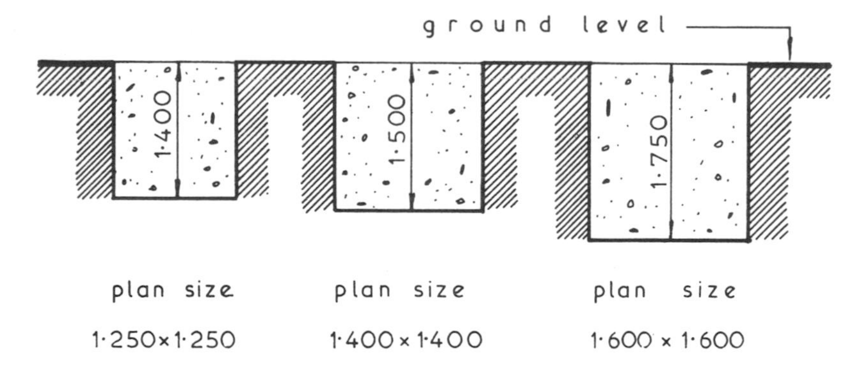

Fig. 8.2.12 Sections through foundation bases

Timesing	Dimension	Squaring	Description
	2		Galvanized steel single-seal medium duty manhole cover and frame to BS 497 size 600 X 450 mm including bedding pointing and flaunching in cement mortar (1 : 3)

Some work is difficult to measure in terms of linear, square or cubic metre Such work is measured by the numbe required and are called enumerated items.

Example of an enumerated item, mea two manhole covers are required.

Work measured as enumerated items includes:

1. Pre-cast concrete units.
2. Air bricks.
3. Chimney pots.
4. Doors.
5. Ironmongery.
6. Drain fittings.

Fig. 8.2.13 Enumerated items

Dimension	Squaring	Description
2.00 4.00	8.00	Thermoplastic floor tiles as before described
7.00 1.00 0.50	3.50	Excavate foundation trench exceeding 0.30 wide commencing at surface strip level maximum depth not exceeding 1.00 m
4.00 1.00 0.75	3.00	
8.00 1.00 1.00	8.00	
	14.50	
2.00 3.00	6.00	Thermoplastic floor tiles as before described
4.00 5.00	20.00	
7.00 3.00	21.00	47.00 m^2

Squaring is the process of multiplying out each set of dimensions to arrive at the total quantity. The answer is then placed opposite the squared dimensio in the squaring column.

This indicates the squaring up of a sin dimension.

Example of squaring several cubic dimensions having the same descriptio Note the total is then placed in a box drawn in the squaring column.

An alternative method is to total the squared dimensions under a line draw at the bottom of the description.

Many surveyors prefer to square figure in a different colour to make them mc legible.

Fig. 8.2.14 Squaring dimensions

Timesing column

Often in building work dimensions repeat themselves. For example Fig. 8.2.15 shows a plan of a building. It can be clearly seen that three of the rooms are 3.00 m × 3.00 m in size and the other four are 4.00 m × 3.00 m. If we were measuring the floor areas the calculations would be as follows:

ROOMS A, B and C	$3.00 \times 3.00 \times 3 = 27\ m^2$
ROOMS D, E, F and G	$4.00 \times 3.00 \times 4 = 48\ m^2$

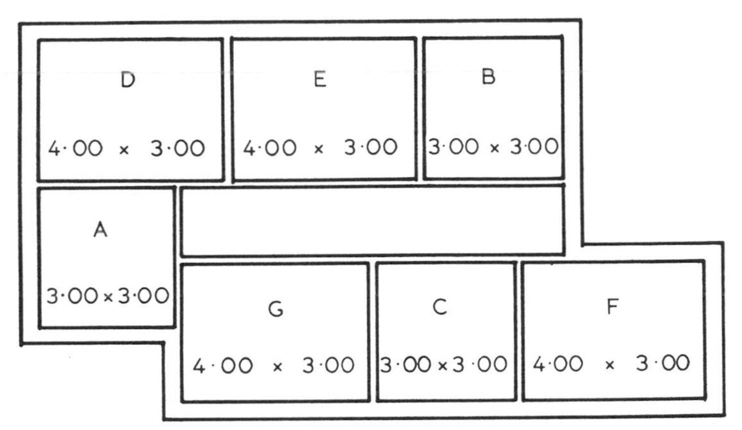

Fig. 8.2.15 Plan of building

The same dimensions can be set down on dimension paper by making use of the timesing column. The figures in the dimension column are simply multiplied by those in the timesing column. To write the above calculation on dimension paper we should proceed as shown in Fig. 8.2.16.

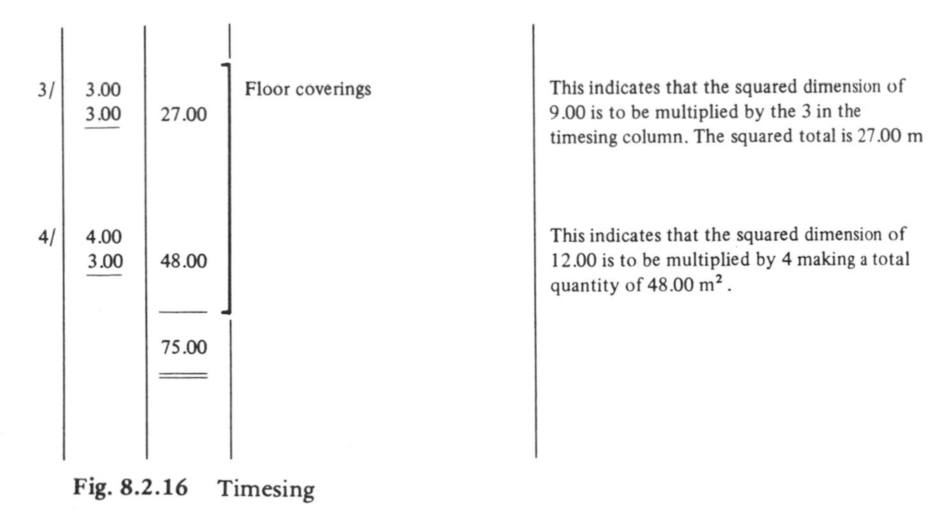

Fig. 8.2.16 Timesing

Timesing can be taken a stage further. Let us assume that the same building is to be three storeys high and that each floor layout is identical. By writing a further figure 3 in the timesing column followed by an oblique line we can indicate that the total number of rooms is 3 × 3 = 9. This is shown in Fig. 8.2.17.

3/3/	3.00 3.00	81.00	Floor coverings	The squared dimension of 9.00 is multiplied by 3 × 3 making a total of 81.00 m².
3/4/	4.00 3.00	144.00		The squared dimension of 12.00 is multiplied by 3 × 4 making a total of 144.00 m².
		255.00		

Fig. 8.2.17

When using the timesing column the taker-off should try to represent the dimensions as closely as possible to those on the drawings. Hence timesing 3 × 3 gives a clearer indication of the number of rooms and storeys than simply timesing by 9.

Although square dimensions are given here as examples the timesing process can be applied in the same way to linear or cubic dimensions and to enumerated items.

Dotting on

When timesing dimensions the taker-off will on occasions need to add. Again the timesing column is used, but this time a dot is used to indicate addition – see Fig. 8.2.18.

The ampersand (&)

When measuring work certain descriptions will have the same-sized dimensions. The taker-off may use the ampersand sign to link the descriptions to the same dimension to save writing the dimension again. The ampersand sign is commonly known as 'anding on' and is written in the description column as & – see Fig. 8.2.19. Several descriptions can be 'anded on' in this manner as Fig. 8.2.20 clearly illustrates.

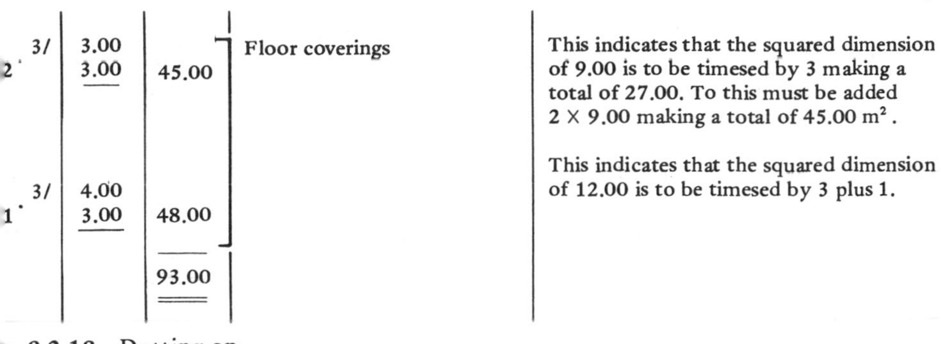

Timesing	Dimension	Squaring	Description	Explanation
3/	3.00		Floor coverings	This indicates that the squared dimension of 9.00 is to be timesed by 3 making a total of 27.00. To this must be added 2 × 9.00 making a total of 45.00 m².
2˙	3.00	45.00		
3/	4.00			This indicates that the squared dimension of 12.00 is to be timesed by 3 plus 1.
1˙	3.00	48.00		
		93.00		

g. 8.2.18 Dotting on

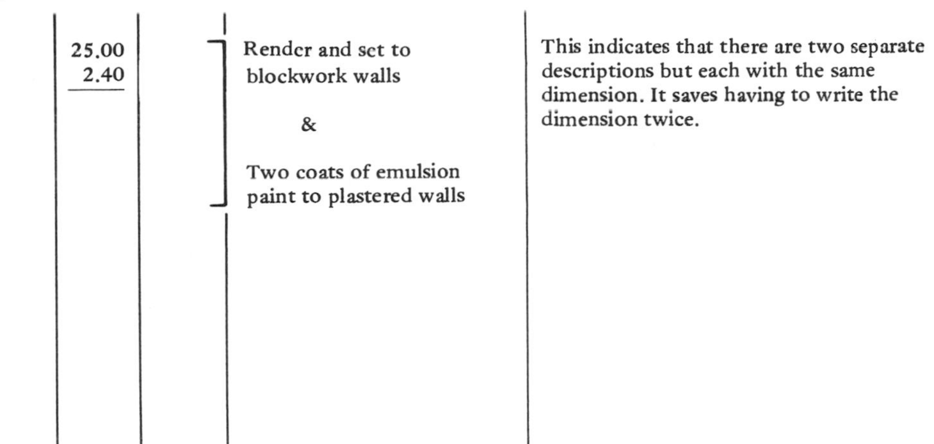

Timesing	Dimension	Squaring	Description	Explanation
	25.00		Render and set to blockwork walls	This indicates that there are two separate descriptions but each with the same dimension. It saves having to write the dimension twice.
	2.40		&	
			Two coats of emulsion paint to plastered walls	

g. 8.2.19 The ampersand

Timesing	Dimension	Squaring	Description	Explanation
	20.00		Hardcore filling average 150 mm thick in making up levels	As above but with three separate descriptions.
	10.00		&	
			Level and compact ground under	
			&	
			Blind surface of hardcore with sand	

ig. 8.2.20

Errors will be made in taking-off quantities as with any other type of work. Where alterations have to be made to dimensions it is better to remove the original dimension and write it out again in as neat a manner as possible — see Fig. 8.2.21.

	~~10.00~~ ~~2.00~~	NIL	Hardcore filling in making up levels average 150 mm thick	The incorrect dimension has the wo 'NIL' written in the squaring colum beside it. The dimension may also be lined through if required.
	10.00 2.50			

Fig. 8.2.21

Waste calculations

Working drawings often do not show the specific dimensions that the taker-off requires. The taker-off must then resort to calculation to find the desired dimensions. Such calculations are called waste calculations. The following points should be observed when writing waste calculations:

1. they should be written as far as possible in the description column of the dimension paper but may spread over the other columns when necessary;
2. they must be set down clearly and logically. Mental arithmetic should not be used and each stage should be capable of checking for accuracy;
3. waste calculations are normally written preceding the description to which they refer. Sometimes, however, it is convenient to calculate several of the waste calculations required at the commencement of taking-off.

An example of waste calculation is given in Fig. 8.2.22.

Order of taking-off

This should follow the order of the SMM.

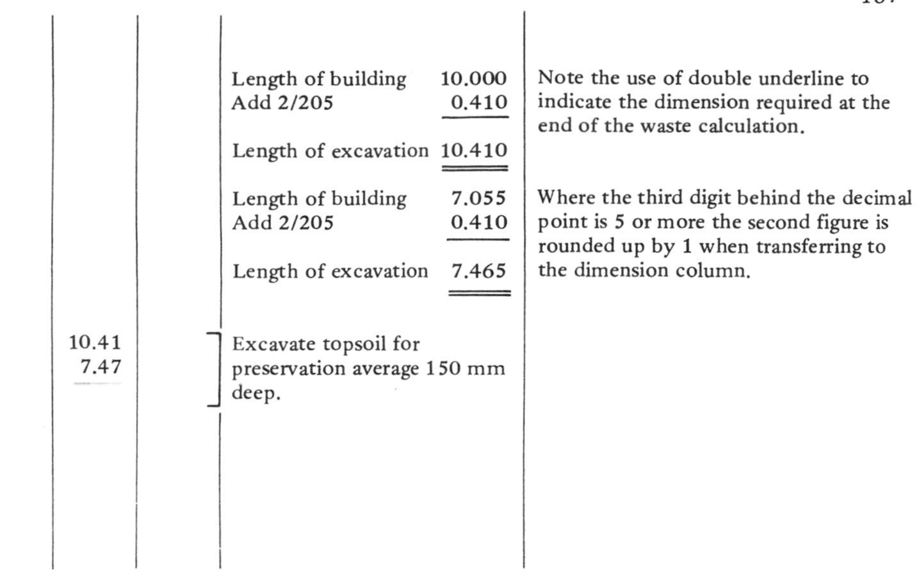

Timesing	Dimensions	Squaring	Description	Notes
			Length of building 10.000 Add 2/205 0.410 Length of excavation 10.410	Note the use of double underline to indicate the dimension required at the end of the waste calculation.
			Length of building 7.055 Add 2/205 0.410 Length of excavation 7.465	Where the third digit behind the decimal point is 5 or more the second figure is rounded up by 1 when transferring to the dimension column.
	10.41 7.47		Excavate topsoil for preservation average 150 mm deep.	

. 8.2.22 Waste calculation

8.3 Mensuration for dimensions

Introduction

In this chapter methods outlined in Section 8.2 are applied to simple mensurational problems including use of:

1. Geometric area and volume formulae.
2. Gradients.
3. Compensation lines.
4. Pythagoras's theorem.
5. Elementary trigonometrical ratios.

Items 4 and 5 are used to calculate rafter lengths, etc.

The same methods are applied to common mensurational problems found in building measurement, namely:

6. Measurement overall and adjustment of wants and voids.
7. Centre-lines and corner adjustments.
8. Interpolation of levels.

Geometric area and volume formulae

In Section 8.2 we studied the basic techniques and terms used in the taking-off process. When measuring we often make use of geometric area and volume formulae. Figures 8.3.1 and 8.3.2 show some of the more common area and volume shapes found in simple building work. The student should first study the shape and the formula and then note the manner in which it is

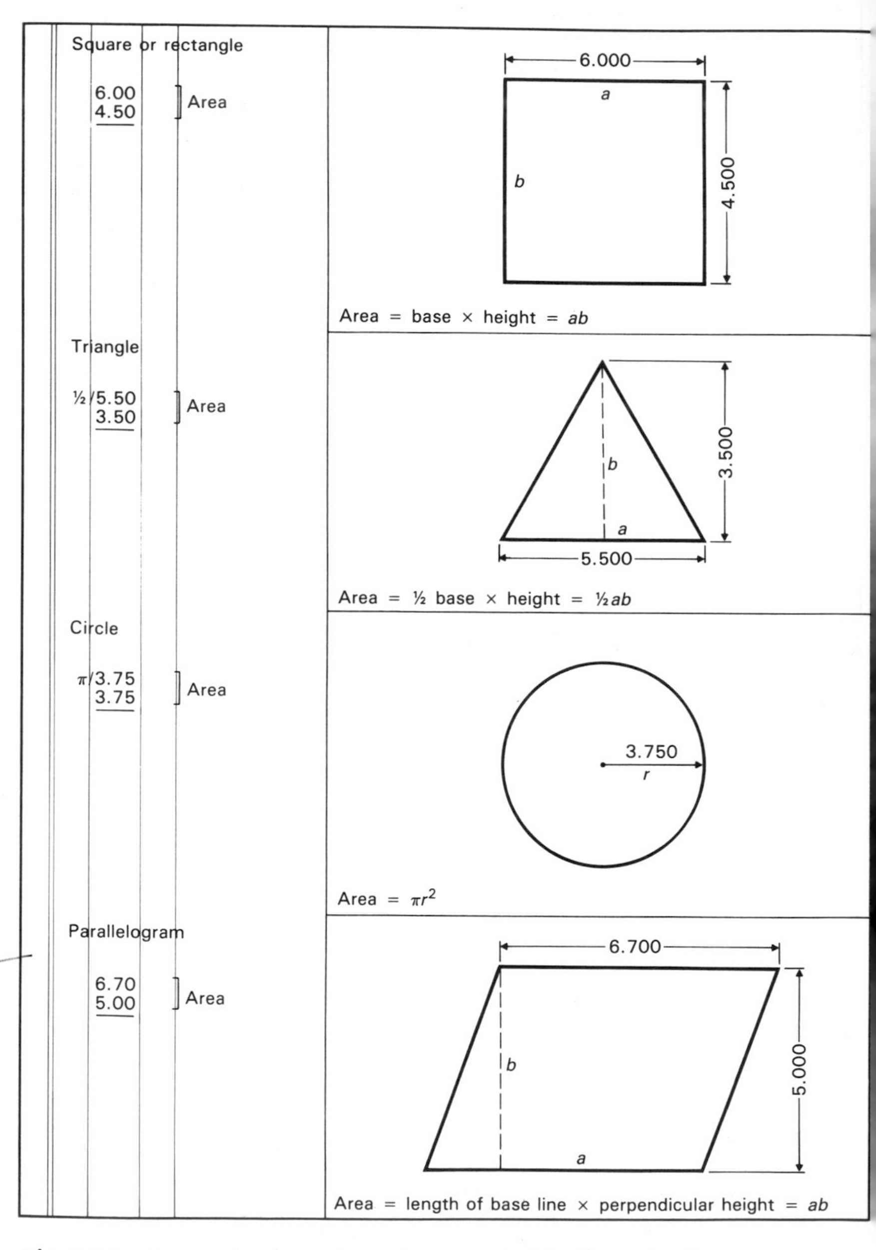

Fig. 8.3.1 Area and volume formulae expressed in dimension format

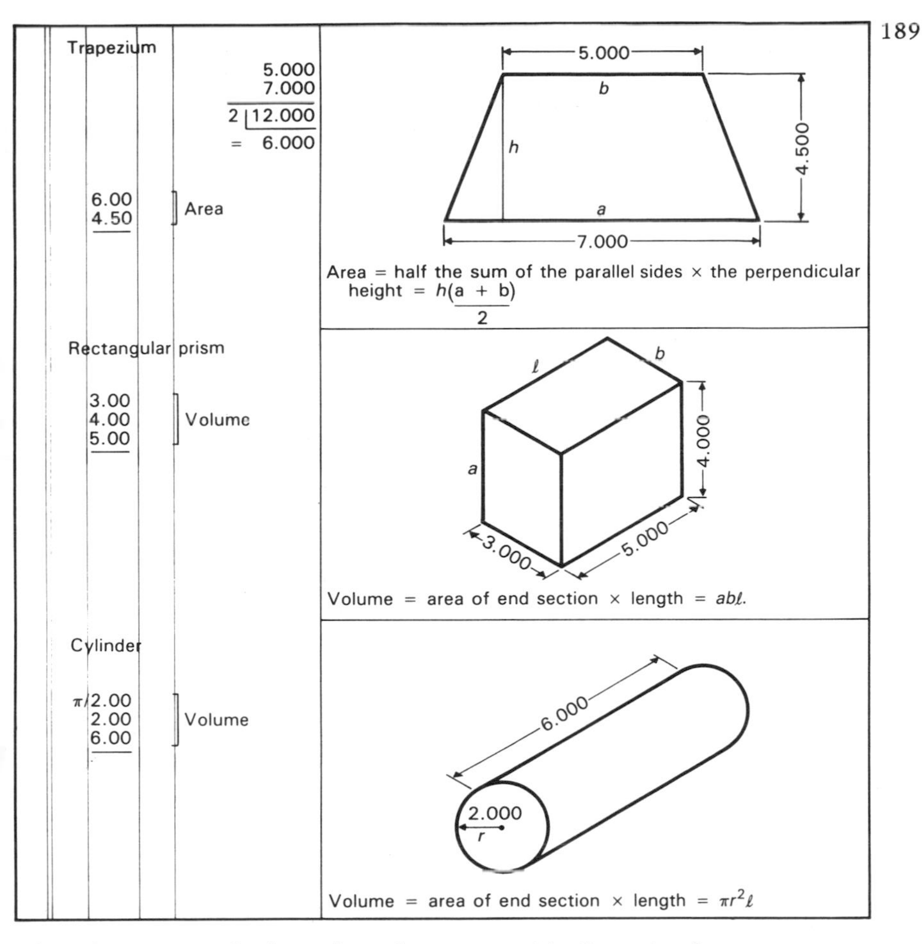

Fig. 8.3.2 Area and volume formulae expressed in dimension format

written on to dimension paper. It is important that the correct sequence of measurement is followed:

linear: length
square : length × breadth or height
cubic: length × breadth × depth

Gradients

A required surface fall or slope in building work may be specified as a gradient. A gradient of 1 in 10 simply means that there is a vertical rise or fall of 1 unit in a horizontal length of 10 units (see Fig. 8.3.3). In certain measurement problems it may be necessary to calculate the length of the

1

10

Fig. 8.3.3 Gradient of 1 in 10

surface slope given the gradient required. Sometimes we may have to incorporate our knowledge of gradients into problems involving the calculation of depths of sewers and the like. Two examples of the use of gradient calculation in measurement are now given.

Example 1 Figure 8.3.4 details a section through an embankment to the side of a motorway. Assuming that the sloping surface c is to be seeded then we need to calculate that length in order to measure the area. The

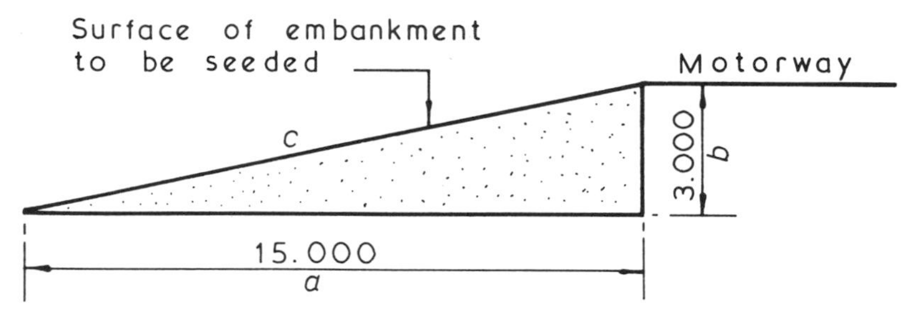

Fig. 8.3.4 Section through embankment

required gradient is 1 in 5. The height of the motorway b is 3.00 m and as the gradient is 1 in 5 the base line a of the triangle so formed must be 15.00 m. Probably the simplest way, then, to calculate length c is to use Pythagoras:

$$\text{Length } c = \sqrt{(a^2 + b^2)}$$
$$\text{Length } c = \sqrt{(225 + 9)}$$
$$\text{Length } c = \sqrt{(234)}$$
$$\text{Length } c = 15.30$$

The length of surface slope for seeding is therefore 15.30 m.

Example 2 Figure 8.3.5 shows a detail of a sewer excavation for a drain which is to be laid at a gradient of 1 in 40. For measurement purposes we need to know the following:

1. The actual depths of all manholes (MH). In this case the invert level of MH A which is existing is 99.000. Therefore, we need to calculate the depth of MH B only.
2. Sewer trench excavation is measured in linear metres stating the average depth. We require, therefore, to calculate the average depth of trench excavation between MH A and MH B.

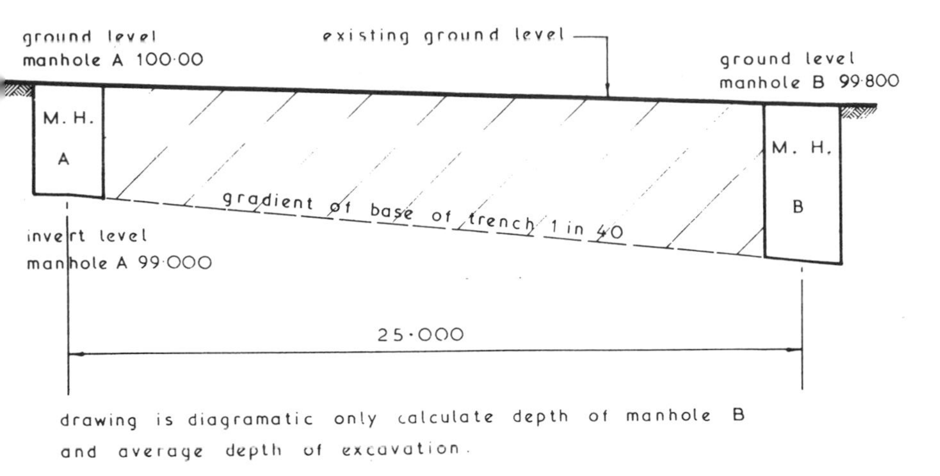

Fig. 8.3.5 Sewer gradients

We may proceed as follows. The depth of MH A and the gradient of the excavation is given.

To find depth of MH B

Fall between MH A and MH B = length of trench × ratio of gradient
= 25.00 × 1/40 = 0.625 m.

Invert level MH A (from drawing)		99.000
Deduct fall between MH A and MH B	=	0.625
Invert level MH B	=	98.375
Depth of MH B		
Ground level MH B (from drawing)		99.800
Invert level MH B		98.375
Depth of MH B	=	1.425 m

Having calculated the depth of MH B it is now simple to find the average depth of trench between MH A and MH B.

Depth of MH A = (100.000 – 99.000)	=	1.000
Depth of MH B	=	1.425
		2.425 m

$$\text{Average depth of trench} = \frac{2.425}{2} = \underline{1.213 \text{ m}}$$

It is common practice in taking-off to use a schedule to calculate all relevant invert levels and average depths of sewer trenches before setting down the totals on dimension paper.

To find the area of a composite shape such as a building site we may proceed by dividing the area into triangles. The total area can then be found by calculating the area of each individual triangle using $\frac{1}{2}$ base × height. For example the area of the triangle ABE in Fig. 8.3.6 would be:

$$\text{Area ABE} = \tfrac{1}{2}(30 \times 10)$$

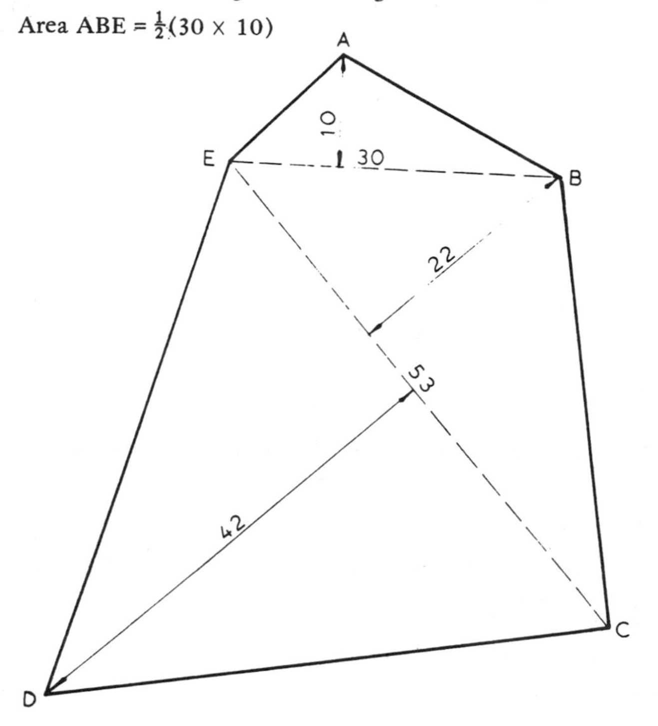

Fig. 8.3.6 Plan of building site

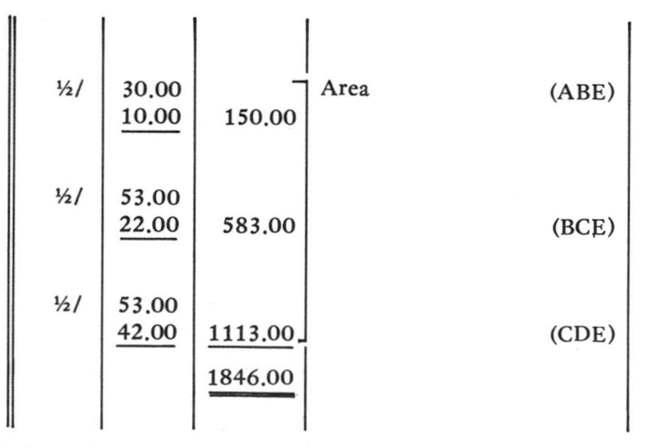

½/	30.00		Area	(ABE)
	10.00	150.00		
½/	53.00			
	22.00	583.00		(BCE)
½/	53.00			
	42.00	1113.00		(CDE)
		1846.00		

Fig. 8.3.7

The total area of the site in Fig. 8.3.6 may be set down on dimension paper as in Fig. 8.3.7.

Figure 8.3.8 indicates a plan of a building site which is irregular and also has one boundary of a curved nature. The quantity surveyor can still make use of triangulation to calculate the area of the site by drawing a compensation line through the curved boundary, in effect making it straight. Compensation lines, often known as give and take lines for obvious reasons, should be drawn as accurately as possible. The area of Fig. 8.3.8 would be set down on dimension paper as shown in Fig. 8.3.9.

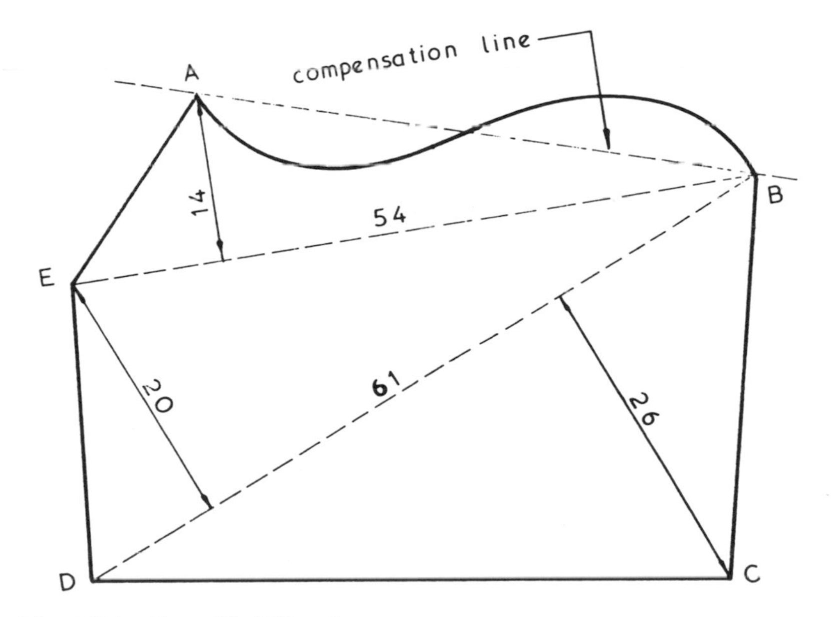

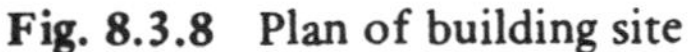
Fig. 8.3.8 Plan of building site

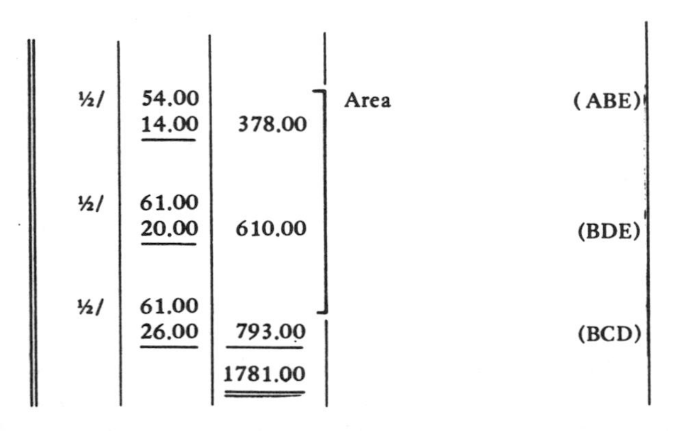

½/	54.00 14.00	378.00	Area (ABE)
½/	61.00 20.00	610.00	(BDE)
½/	61.00 26.00	793.00	(BCD)
		1781.00	

Fig. 8.3.9

Wants and voids

These two terms are used in measurement work in the following manner.

Wants

A want is an opening or a break along the boundary of an area, room, wall, etc. being measured. Wants are always deducted whatever their size. Figure 8.3.10 illustrates typical wants.

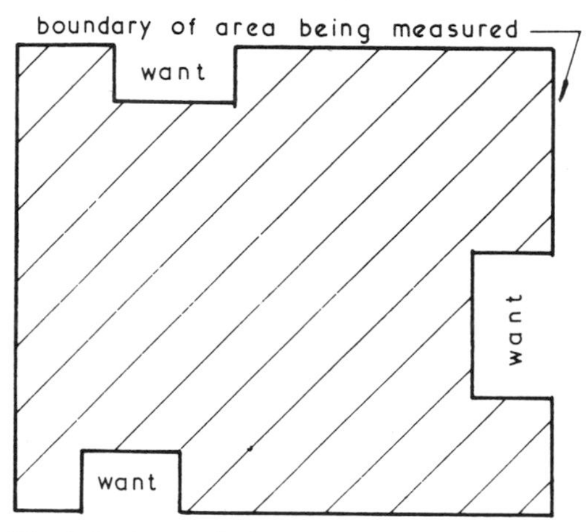

Fig. 8.3.10 Typical wants

Voids

A void differs from a want in that it is wholly within and detached from the boundaries of the area being measured. Voids are not always deducted when measuring for Bills of Quantities. The SMM lays down the minimum size a void must be before any deduction is made to the area being measured. For example:

1. No deduction in brickwork shall be made for a void not exceeding 0.10 m^2.
2. No deduction on formwork shall be made for voids not exceeding 5.00 m^2.
3. No deduction of timber boarding and flooring shall be made for voids not exceeding 0.50 m^2.

Figure 8.3.11 illustrates typical voids.

Centre-line calculation

Centre-lines are used extensively in the measurement of the structure of a building. Examples are excavation of foundation trenches, backfilling to trenches, temporary support to foundation trenches, concrete in foundations, brickwork and blockwork. Use will be made of centre-line calculations in the worked measurement exercises later in the book. It is important, therefore,

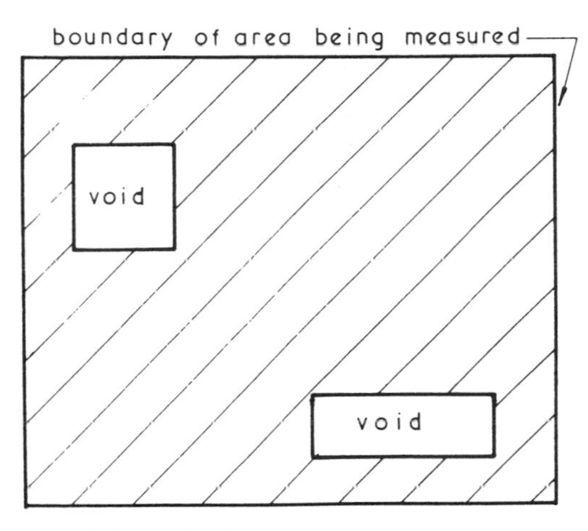

Fig. 8.3.11 Typical voids

that the student be able to calculate centre-lines accurately. It is fair to say that many students find it difficult to do so initially and for this reason the following pages contain several worked examples set out on a stage-by-stage basis.

What are centre-lines?

What exactly do we mean by a centre-line and how do we set about calculating them? Figure 8.3.12 details the plan view of a brick wall. To measure its length accurately we need to calculate the length of the centre-line of the wall. This is because if the wall were laid out in a straight line its length would be equal to a line measured through the centre of the wall.

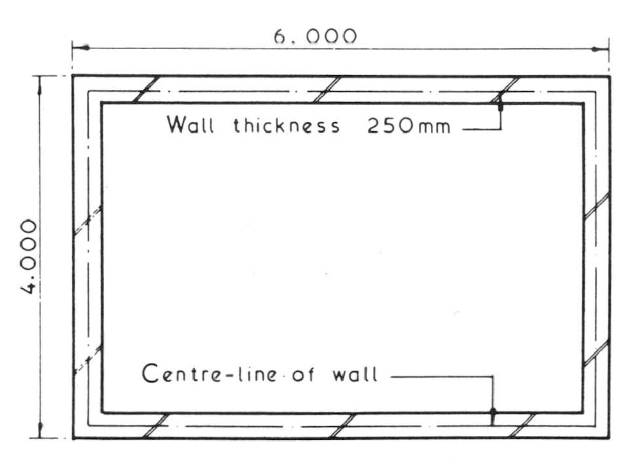

Fig. 8.3.12 Centre-line calculation

 To calculate centre-line of Fig. 8.3.12

The external perimeter length of brickwork from drawing = 2/6.000 12.000
= 2/4.000 8.000

20.000

Consider plan of one corner Fig. 8.3.13. Clearly at this one corner the perimeter of the brickwork is longer than the centre-line of the wall by $A + B$.

There are four corners in this wall so if we were to deduct length $A + B \times$ the four corners from the perimeter of the brick wall we would have calculated the length of the centre-line.

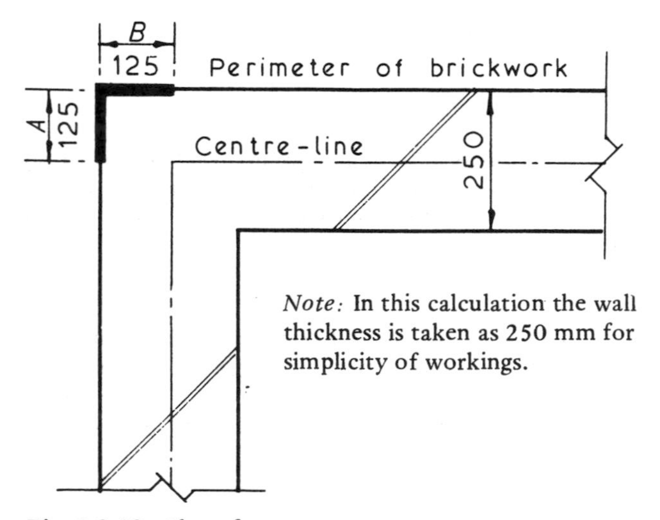

Fig. 8.3.13 Plan of one corner

The wall thickness is given as 250 mm wide. Both dimensions A and B must equal half the total wall thickness. Centre-line of wall:

Perimeter of brickwork as before	=	20.000
Deduct		
4/2/125 (2 × 125 per corner)	=	1.000
and four corners		
centre-line of wall	=	19.000 m

Example 3 Figure 8.3.14 shows a plan and section of a trench fill type foundation. Unless otherwise specified it is always assumed that the wall is central of the foundation and therefore the length of the centre-line of the brick wall and the foundation trench and concrete are the same. The wall size of 250 mm wide, although non-standard, is chosen for simplicity of calculation in this first example.

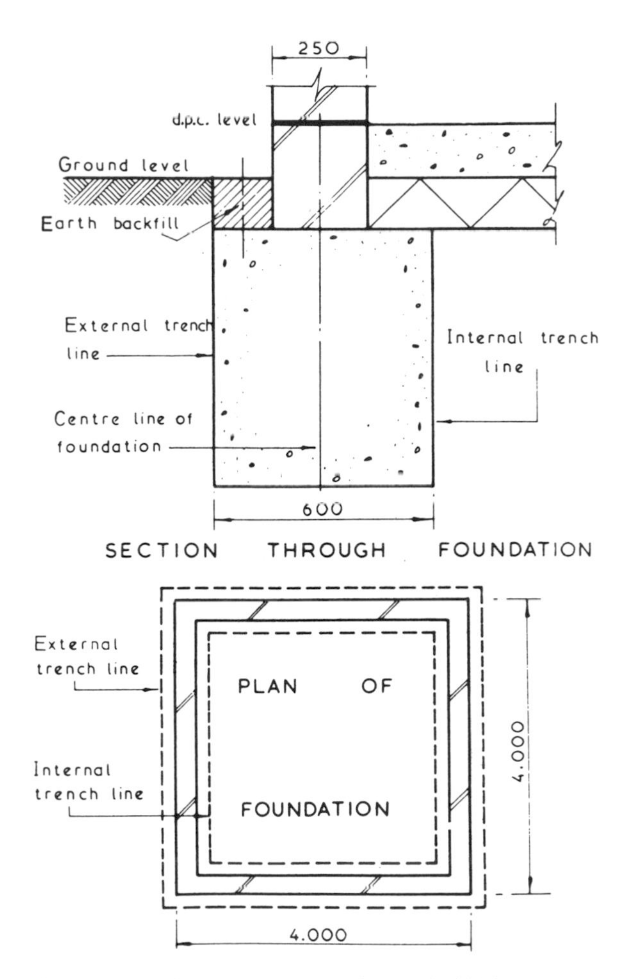

Fig. 8.3.14 Plan and section of trench fill foundation

The following lengths may be required for measurement purposes:

1. Centre-line of the brick wall and foundation.
2. Length of external trench line.
3. Length of internal trench line.
4. Centre-line of earth backfilling.

1. To calculate centre-line of brick wall and foundation

Consider Fig. 8.3.15.

Perimeter of brickwork from plan	= 4/4.000	=	16.000
Deduct 4/2/125		=	1.000
Centre-line length	=		15.000

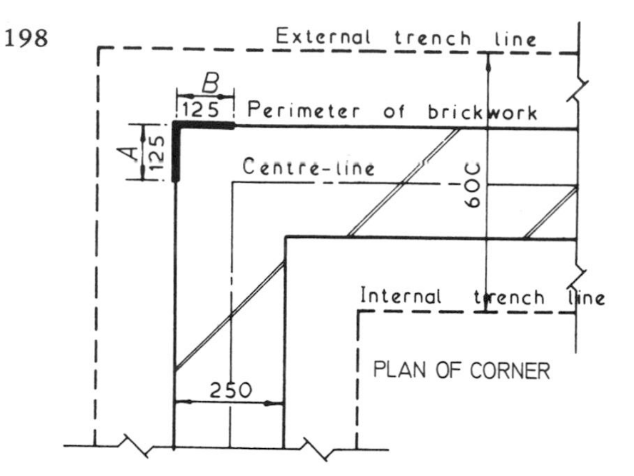

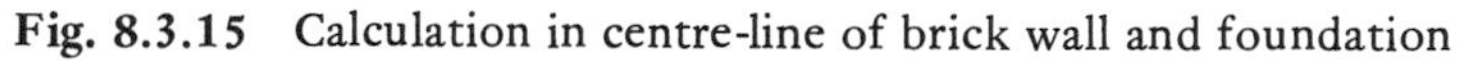

Fig. 8.3.15 Calculation in centre-line of brick wall and foundation

2. To calculate length of external trench line

Consider Fig. 8.3.16. Clearly the length of external trench line is longer than the centre-line by $A + B$ at each corner. External trench line:

Centre-line of foundation as before	=	15.000
Add 4/2/300 (2 × 300 per corner) × 4 corners	=	2.400
length of external trench line	=	17.400

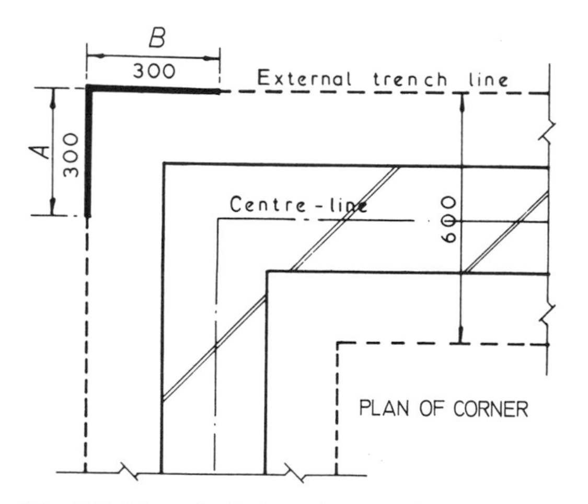

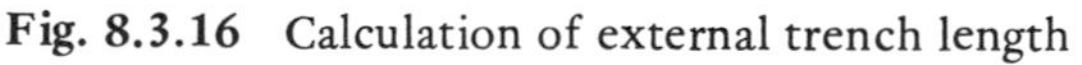

Fig. 8.3.16 Calculation of external trench length

3. Length of internal trench line

Consider Fig. 8.3.17. Clearly this is the opposite situation of the external trench line.

Centre-line as before	=	15.000
Deduct 4/2/300	=	2.400
length of internal trench line		12.600

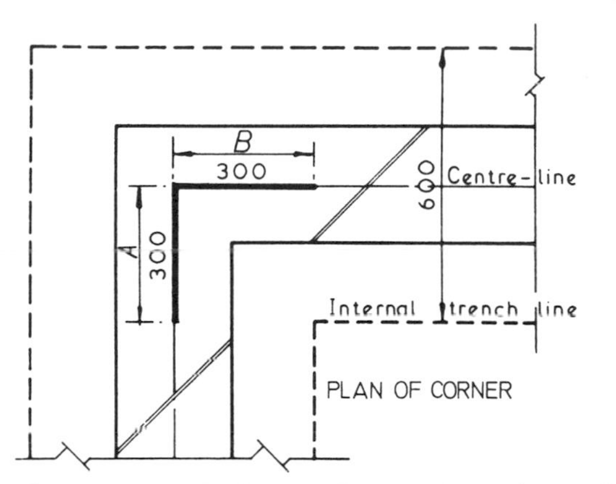

Fig. 8.3.17 Calculation of internal trench length

4. Centre-line of earth backfill

Consider Fig. 8.3.18. The centre-line of the earth backfill is central between the external trench line and the perimeter of the brick wall. Its length, therefore, must be the average of those two.

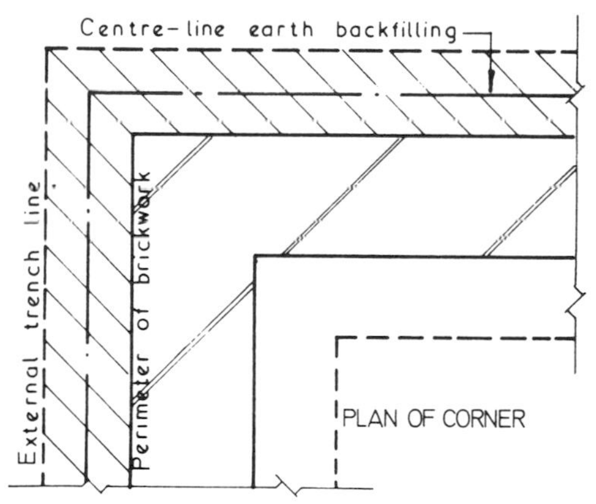

Fig. 8.3.18 Calculation of centre-line of earth backfill

Length of external trench line	=	17.400
Length of perimeter of brickwork	=	16.000
	=	33.400

Centre-line of earth backfill = $\frac{33.400}{2}$ = 16.700

Probably the most difficult centre-lines to calculate are those involving cavity walls where the individual skins are a different thickness. Figure 8.3.19 indicates a section and plan of such a foundation. For simplicity a trench fill style foundation is used.

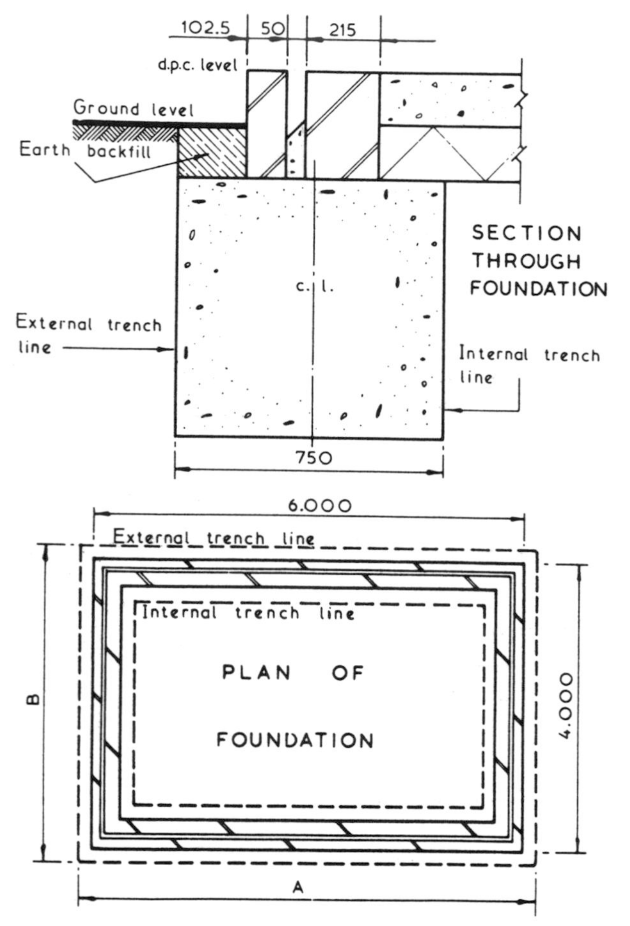

Fig. 8.3.19 Plan and section of trench fill foundation

For measurement purposes we need to calculate the following:

1. Length of external trench line.
2. Centre-line of foundation trench.
3. Length of internal trench line.
4. Centre-line of external brick wall.
5. Centre-line of cavity.
6. Centre-line of internal brick wall.
7. Centre-line of earth backfill.

Example 4 In this example we will use different methods to those used in Example 3 to indicate that alternative methods may be adopted. Again we are to assume that the mass of the whole wall is sitting central on the foundation concrete.

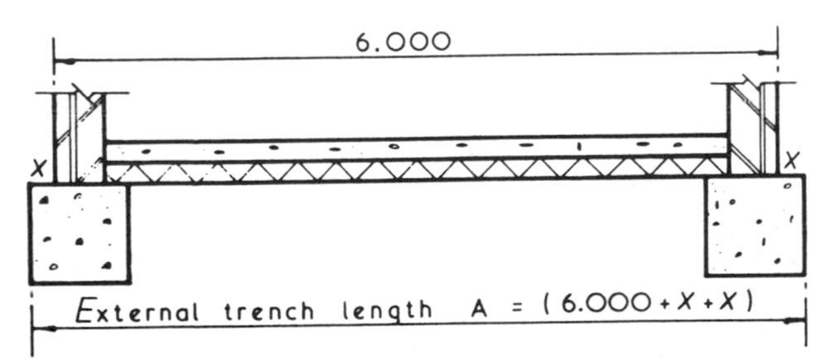

Fig. 8.3.20 Calculation of projection

1. To calculate external trench length

Consider Fig. 8.3.20. If we can find dimension x then clearly we can calculate external trench length for one side. To find x:

Width of foundation	=	750 mm
Width of wall	=	367.5 mm
	2)	382.5 mm
projection of foundatio.	=	191.25 mm

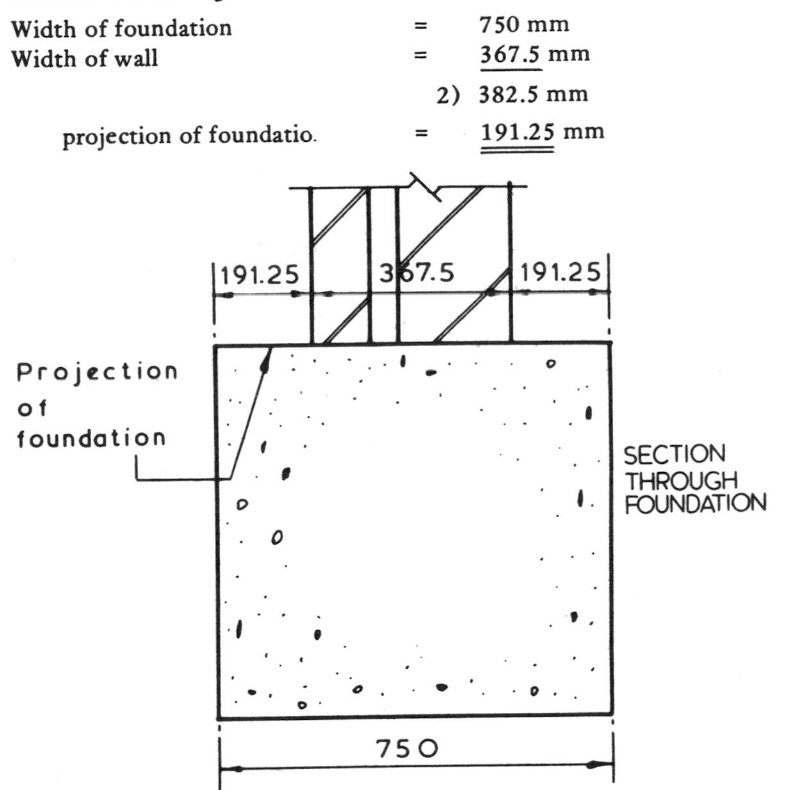

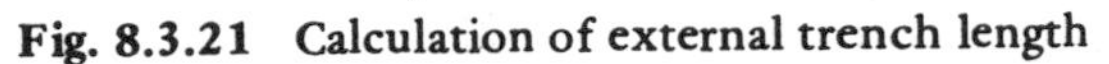

Fig. 8.3.21 Calculation of external trench length

Length of external trench length *A*	= (6.000 + 2/191.25) = 6.383	
Length of external trench length *B*	= (4.000 + 2/191.25) = 4.383	
Total length of external trench	= 2/6.383	= 23.766
	= 2/4.383	= 8.766
		21.532

2. To calculate centre-line of foundation trench

Consider Fig. 8.3.22.

External trench perimeter as before	= 21.532
centre-line of foundation deduct corners 4/2/375	= 3.000
	18.532

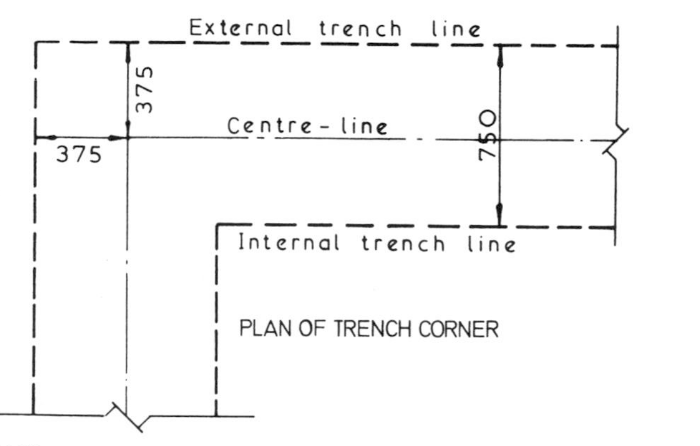

Fig. 8.3.22 Calculation of centre-line of trench

3. To calculate internal trench line

Consider Fig. 8.3.23.

External trench length as before	= 21.532
internal trench length deduct corners 4/2/750	= 6.000
	15.532

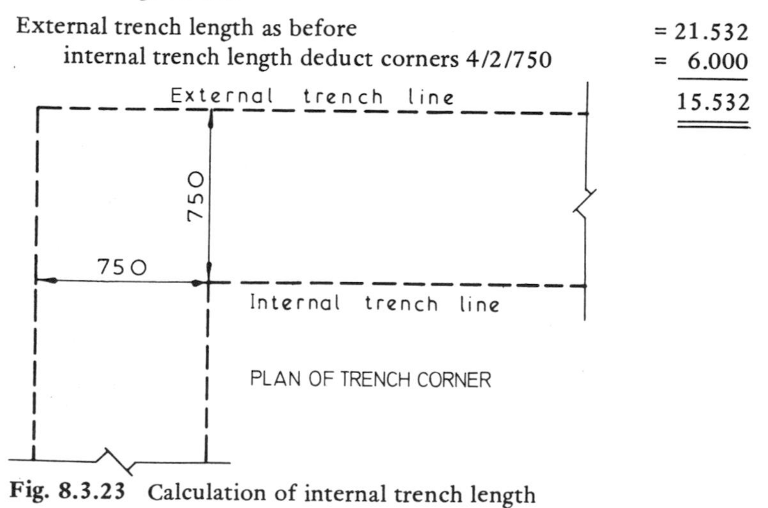

Fig. 8.3.23 Calculation of internal trench length

Wall calculations

Consider Fig. 8.3.24.

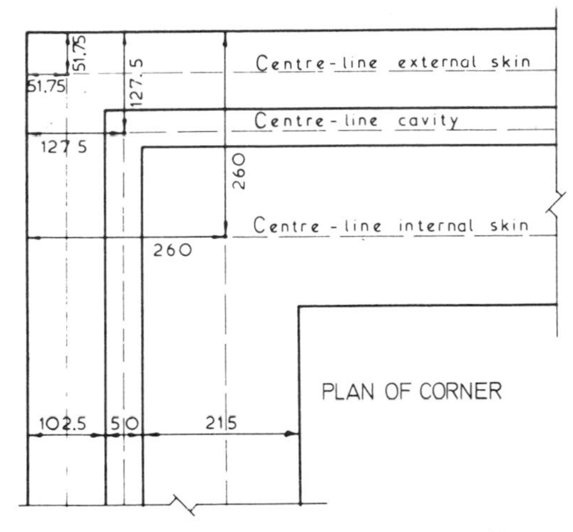

Fig. 8.3.24 Calculation of wall centre-lines

4. *Centre-line external skin*

Perimeter of brickwork 2/6.000 + 2/4.000	=	20.000
Deduct corners 4/2/51.75	=	0.410
	℄	19.590

5. *Centre-line of cavity*

Perimeter of brickwork as before	=	20.000
Deduct corners 4/2/127.5	–	1.020
	℄	18.980

6. *Centre-line of internal skin*

Perimeter of brickwork as before	=	20.000
Deduct corners 4/2/260	=	2.080
	℄	17.920

7. *Centre-line of earth backfill*

External trench length as before	=	21.532
Add perimeter of brickwork	=	20.000
Average	=	2)41.532
	℄	20.766

So far we have considered only those plan shapes that are square or rectangular. Figures 8.3.25 and 8.3.26 indicate two plan shapes that at first sight do not appear to be rectangular. To calculate the centre-lines of either of these buildings we would need to know the following:

1. The length of the external perimeter of the wall.
2. The number of corners that need to be adjusted to calculate centre-line lengths.

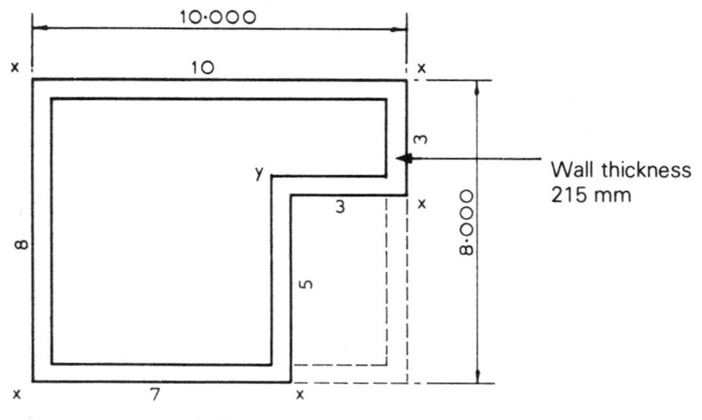

Fig. 8.3.25 Plan of building

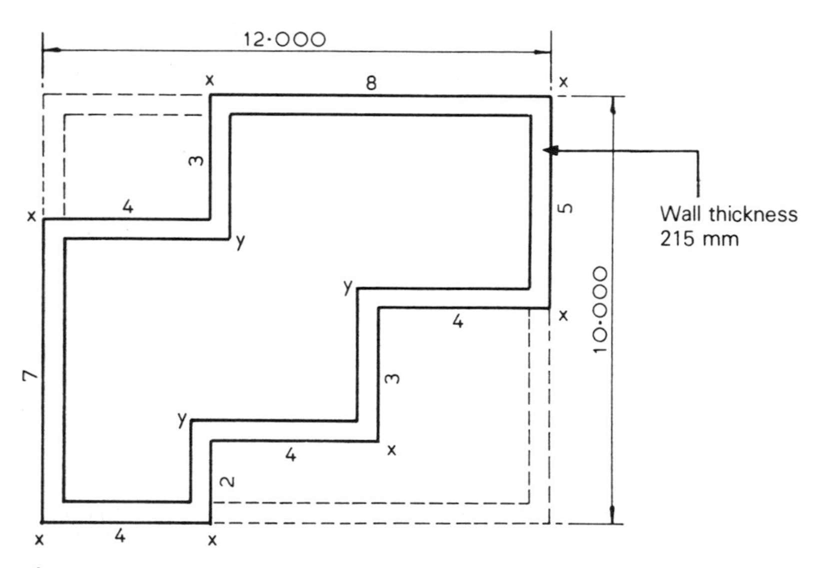

Fig. 8.3.26 Plan of building

Consider Fig. 8.3.25. If we measure around the building in a clockwise direction we can calculate the external perimeter length of the wall:

$$10 + 3 + 3 + 5 + 7 + 8 = 36.000 \text{ m}$$

Alternatively, if we calculate the perimeter of wall by using the overall dimensions we find that we arrive at the same answer:

2/10.000 = 20.000
2/ 8.000 = 16.000

Perimeter = 36.000 m

Clearly, the length of the set back must be the same as the rectangle formed by the dotted line, see Fig. 8.3.27.

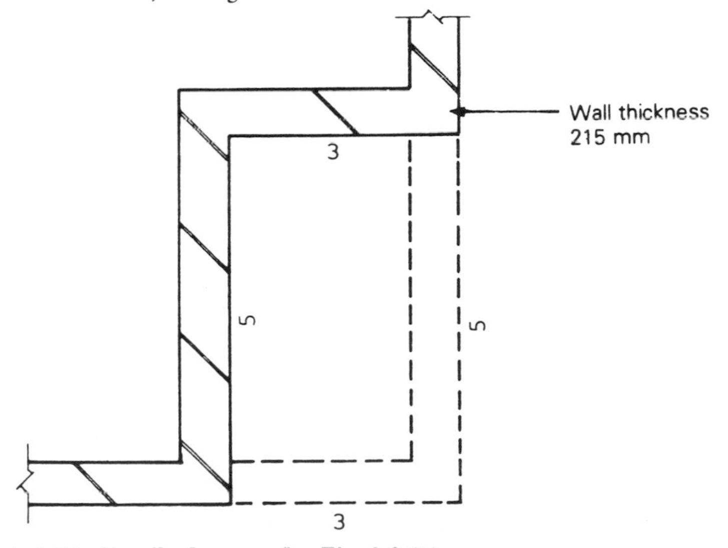

Fig. 8.3.27 Detail of corner for Fig. 8.3.25

Centre-line

Figure 8.3.25 has five corners in total. We have to consider the number of corners that have to be adjusted to calculate the length of centre-line. From Fig. 8.3.25 we see that there are five external corners (x) and one internal corner (y). By deducting the internal corners from the external corners we can arrive at the number of corners to be adjusted. From Fig. 8.3.25:

External corners (x) = 5
Internal corners (y) = 1

Numbers of corners to be adjusted = 4

centre-line = perimeter as before	= 36.000
Deduct corners 4/2/107.5	= 0.860
centre-line	= 35.140

We can therefore confirm the general rule that for plan shapes similar to Figs. 8.3.25 and 8.3.26 we can treat them as rectangles or squares and ignore the set backs. The number of corners to be deducted will always be four where the walls join up to enclose a building.

To prove these observations let us consider Fig. 8.3.26.

Perimeter

Again measuring around building in a clockwise direction:

8 + 5 + 4 + 3 + 4 + 2 + 4 + 7 + 4 + 3 = 44.000 m

Alternatively, using overall dimensions,

	2/12.000 =	24.000
	2/10.000 =	20.000
the same as before		44.000 m

Centre-line

External corners (x)	= 7
Deduct internal corners (y)	= 3
Number of corners to be adjusted	= 4

Perimeter as before	= 44.000
Deduct corners 4/2/107.5	= 0.860
centre-line	= 43.140 m

Projections

Buildings often have projections along the length of a wall. Figure 8.3.28 is a typical example. To arrive at the perimeter of this building we can mentally move the shaded portion of the wall length A into the position shown dotted. This leaves the two lengths shaded black which must be added to arrive at the total perimeter.

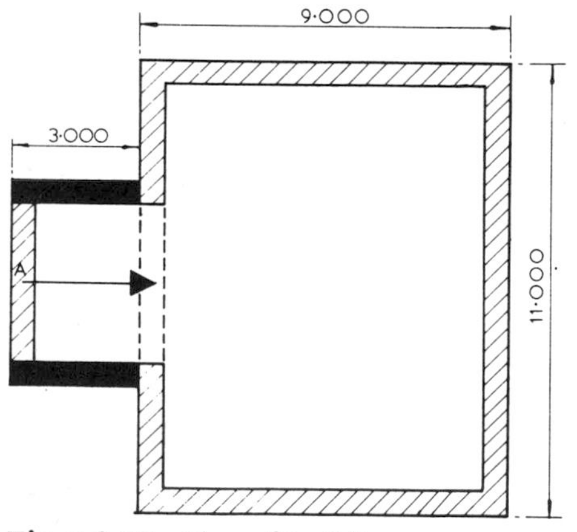

Fig. 8.3.28 Plan of building

For example:

Perimeter =	2/ 9.000	=	18.000
	2/11.000	=	22.000
Add projection	2/ 3.000	=	6.000
External perimeter length of building		=	46.000

Centre-lines should then be calculated in the normal way deducting four corners.

A further example of this type of projection is given in Fig. 8.3.29. Again we can make the shape rectangular by mentally slotting the shaded areas into the dotted positions by following the direction of the arrows. We are then left with the two black areas of wall which must be added to the overall rectangular size.

External wall perimeter =	2/13.000	=	26.000
	2/10.000	=	20.000
Add black shaded areas	2/ 2.000	=	4.000
External perimeter length of building		=	50.000

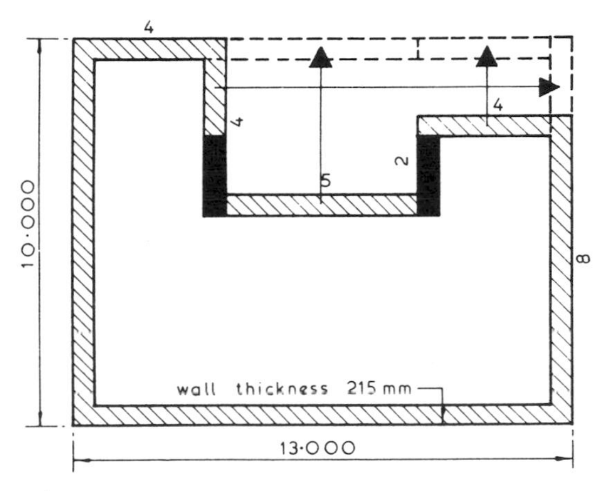

Fig. 8.3.29 Plan of building

Alternatively, we could simply add up the dimensions around the building:

4 + 4 + 5 + 2 + 4 + 8 + 13 + 10 = 50.000 m

Centre-line

External corners	=	6
Internal corners	=	2
numbers of corners to be adjusted	=	4

Students for one reason or another find difficulty generally with centre-line calculation. The author strongly advises students when calculating centre-lines to draw a plan view sketch of the corner on a rough piece of paper, similar to the drawings used in this book. This helps the student to visualise the problem.

Interpolation of levels

It is necessary for students to make use of interpolation of levels in measurement as well as in surveying.

Frequently this will involve the calculation of the volume of excavation over a building site. It is common for building sites to be of a sloping nature and excavation and filling is undertaken to make the site flat. The quantity surveyor has to calculate the amount of excavation and of filling material required. To enable this to be done a grid of regularly spaced spot ground levels is prepared. The level to which the site is to be excavated down to is known as the reduced level. Figure 8.3.30 shows a section through a building site, the reduced level required being 100.00.

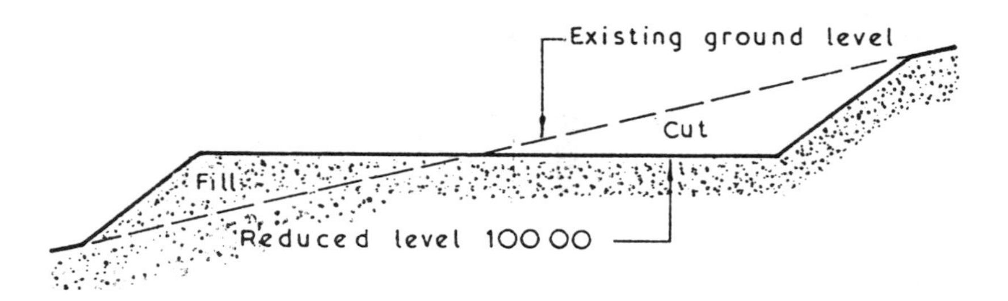

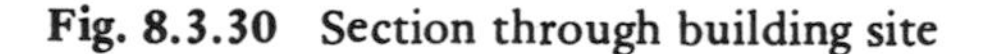

Fig. 8.3.30 Section through building site

Let us assume that a grid of levels has been prepared for a site as shown in Fig. 8.3.31. The reduced level required is 100.00. Soil above the 100.00 line would need to be excavated, and below the 100.00 line fill material would be required.

To measure the quantity of excavation we need to be able to plot the contour line 100.00. Clearly this passes through the middle squares and its position can be calculated from Fig. 8.3.32.

<u>Position of contour line 100.00 on grid line B2–B3</u>

$$\text{Length of grid line} \left(\frac{\text{reduced level} - \text{B3}}{\text{B2} - \text{B3}}\right) =$$

$$= 10 \left(\frac{100.00 - 99.50}{101.00 - 99.50}\right) = 10 \times \frac{0.50}{1.50} = 3.33 \text{ m}$$

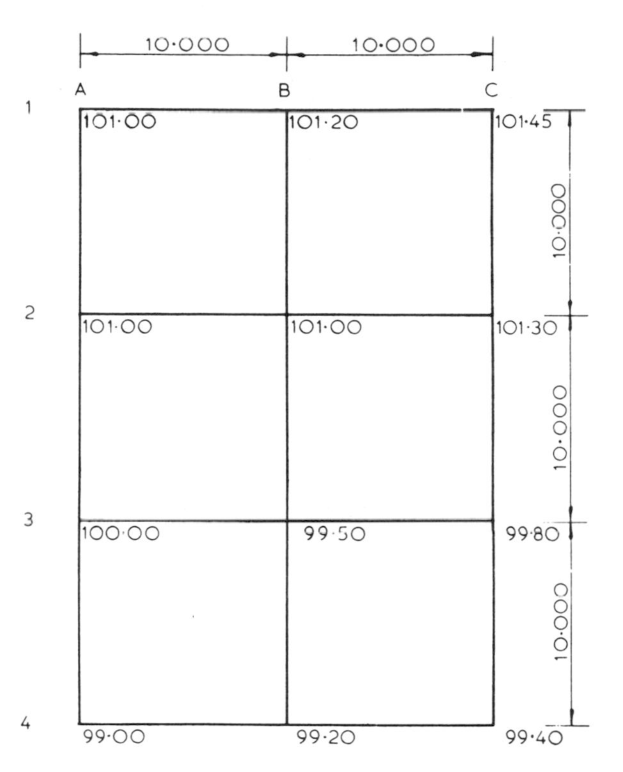

Fig. 8.3.31 Grid of levels

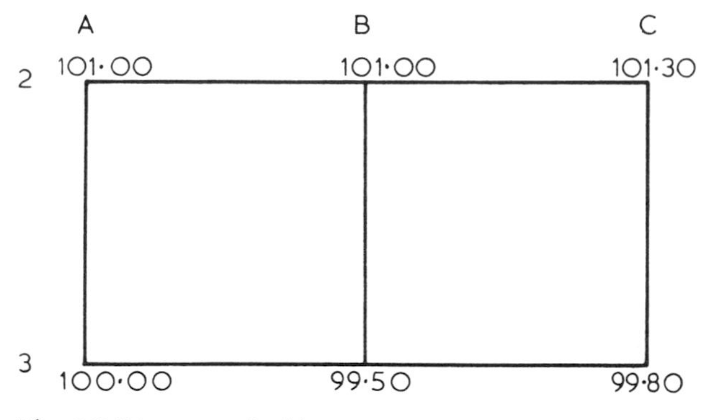

Fig. 8.3.32 Part of grid

$$\frac{\text{Position of contour line 100.00 on grid line C2–C3}}{10} \left(\frac{100.00 - 99.80}{101.30 - 99.80}\right) = 10 \times \frac{0.20}{1.50} = 1.33 \text{ m}$$

We can now plot the reduced level line of 100.00 as in Fig. 8.3.33.

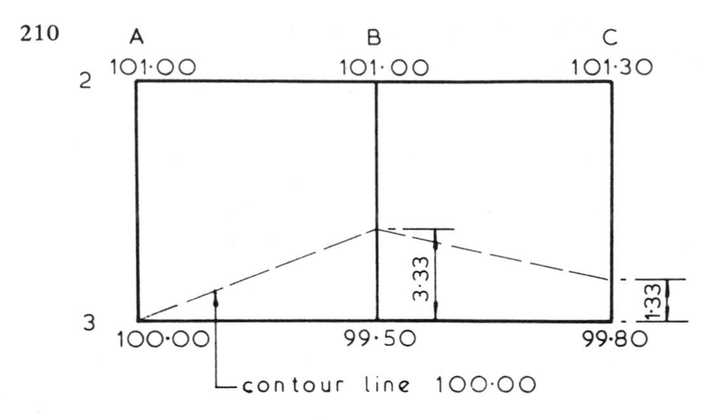

Fig. 8.3.33 Contour line 100.00

It is necessary at this level for the student to be able to interpolate levels only.

8.4 Worked example: Introduction to taking-off dimensions and billing direct

The value of measuring a simple foundation cannot be over-emphasised. Three important sections of the SMM are involved, namely excavation and earthwork, concrete work and brickwork and blockwork. Greater use will also be made of centre-line calculations in this example and one will realise why so much time was spent on learning to calculate these in Section 8.3.

Two alternative methods of measuring the brickwork are given because some students find this particularly difficult.

Figure 8.4.1 details the foundation work which is again of trench fill construction. Reference should also be made to Figs. 8.4.2, 8.4.3 and 8.4.4 which show the sequence of measurement in diagram form.

Specification notes

Topsoil excavation	To be retained on site for later use.
Trench excavation	Excavated material to be removed from site.
Foundation concrete	Plain mass concrete (1:3:6 – 19).
Brickwork	102.5 mm brickwork in cavity construction in calcium silicate bricks class 3 in stretcher bond in cement mortar (1:4). Facing bricks measured to three courses below d.p.c. level. Allow the prime cost sum of £100 per 1000. Pointed with a rubbed-in joint.
Cavity	Form 50 mm wide cavity between skins of brickwork and fill with weak concrete to within 150 mm of d.p.c. level.
Oversite construction	
Hardcore	Clean brick hardcore 150 mm thick well compacted and consolidated and blinded with sand.
Damp-proof membrane	1000 gauge polythene with minimum 200 mm laps laid on blinded hardcore bed.
Concrete bed	Reinforced concrete bed (1:2:4 – 20) 150 mm thick incorporating fabric reinforcement to BS 4483 type B.196 weighing 3.05 kg/m^2 with minimum 150 mm end and side laps.
Damp-proof courses	Hessian-based bituminous felt to BS 743.

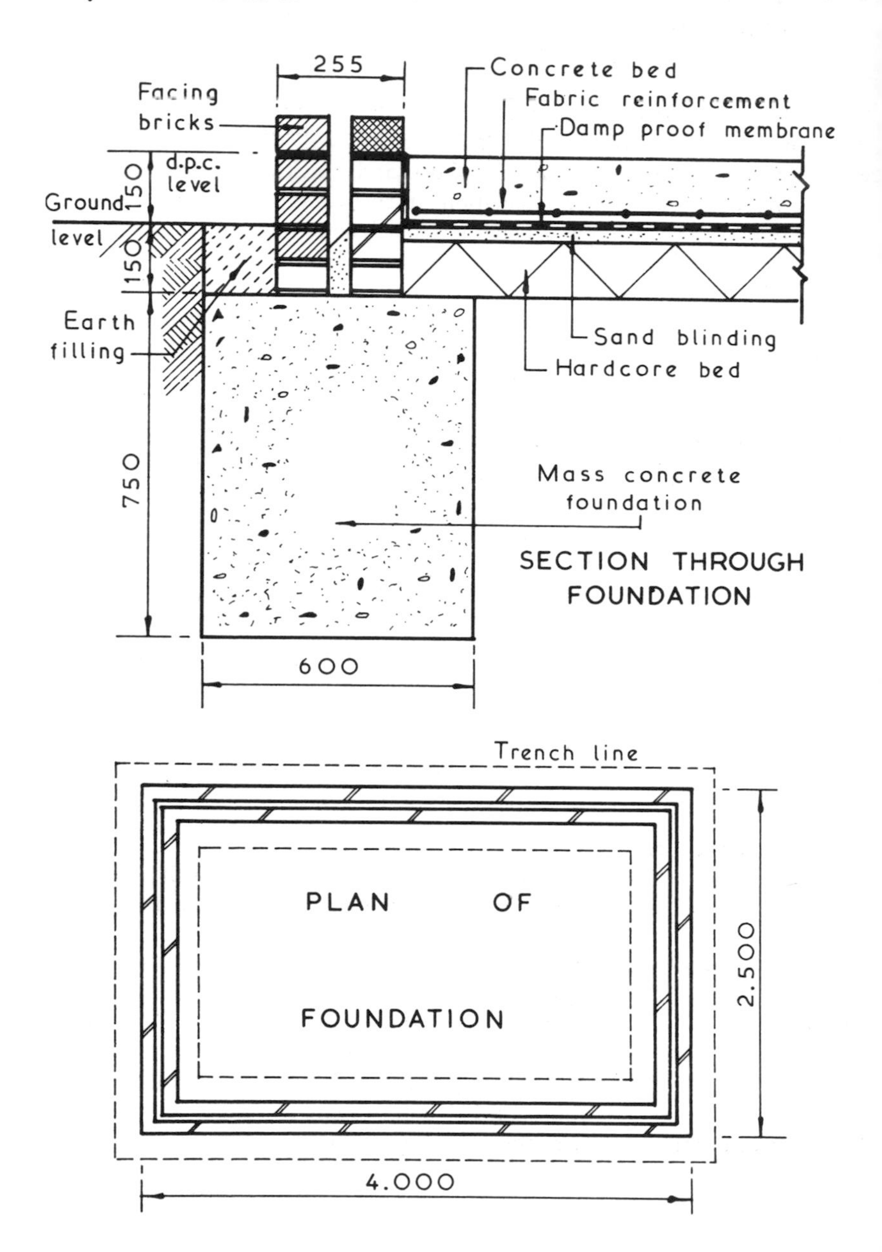

Fig. 8.4.1 (To be read in conjunction with Specification notes shown at the beginning of this section)

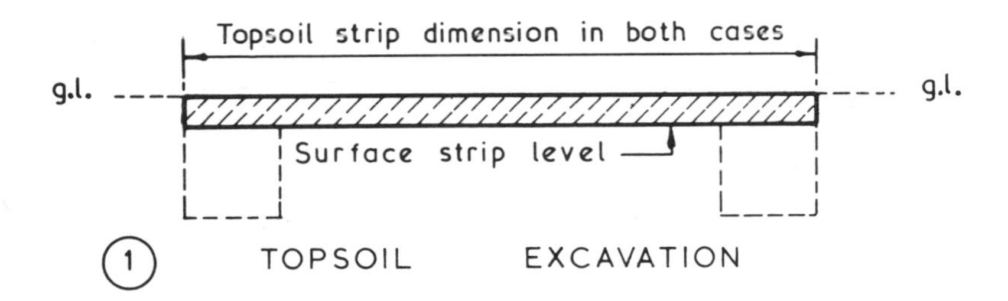

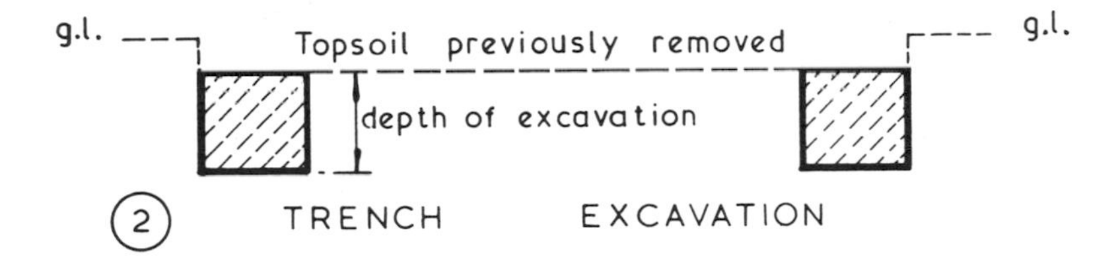

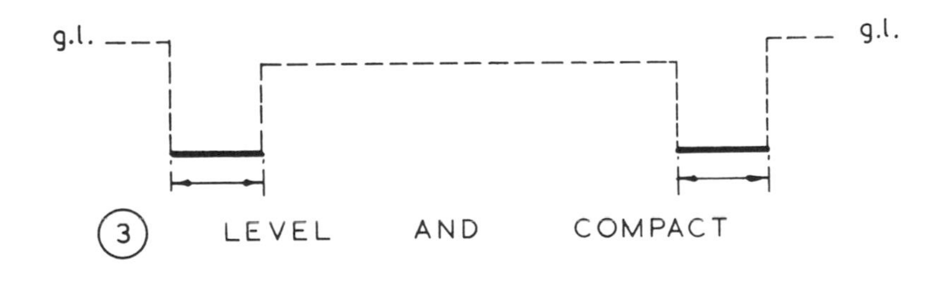

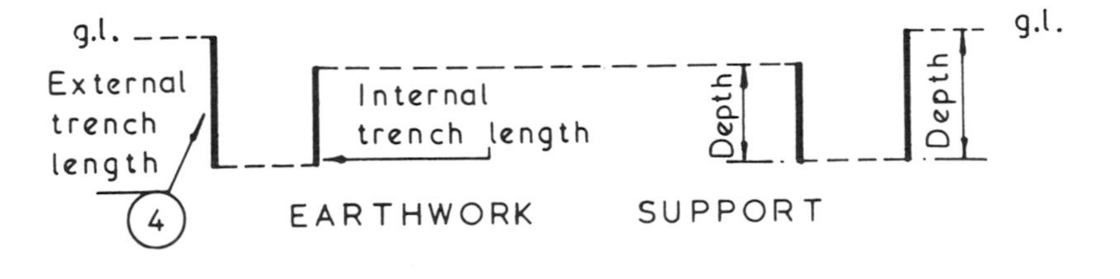

Fig. 8.4.2 Measurement sequence

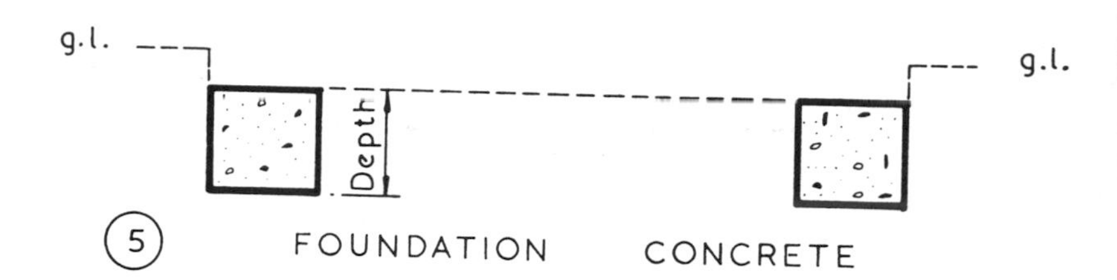

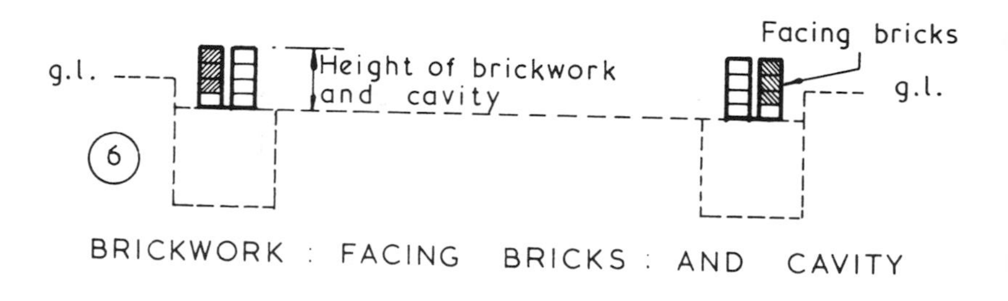

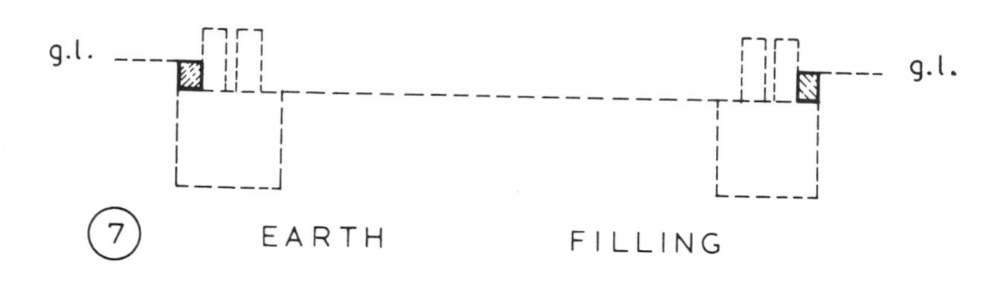

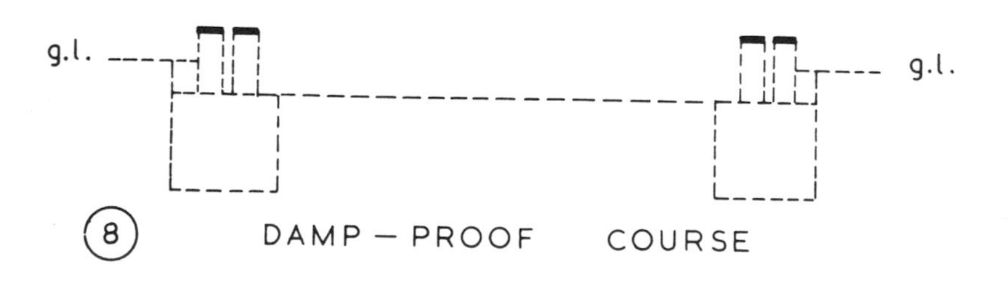

Fig. 8.4.3 Measurement sequence

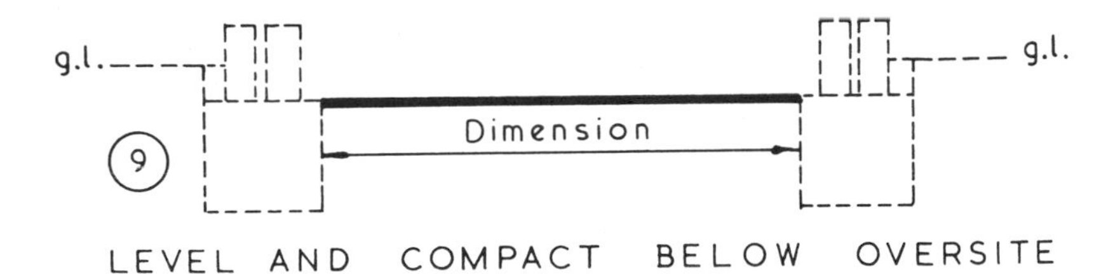

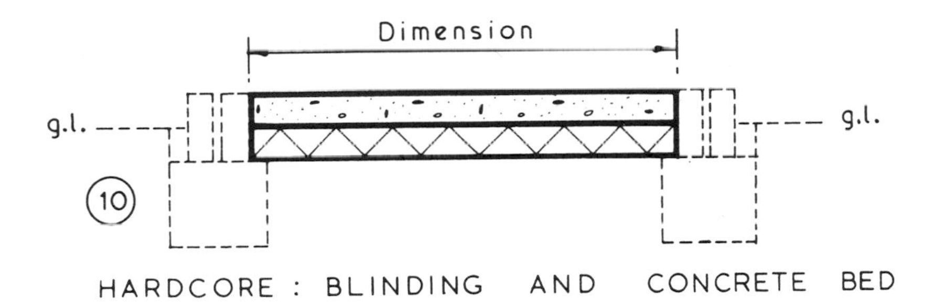

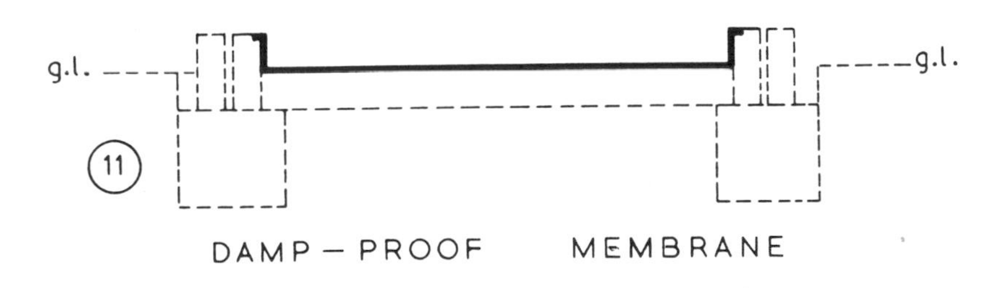

Fig. 8.4.4 Measurement sequence

Take-off list (For figure 8.4.1)

1) Excavate topsoil
2) Deposit topsoil
3) Excavate trench
4) Disposal of excavated material
5) Earthwork support
6) Level and compact base of excavation
7) Foundation concrete
8) Brickwork
9) Facing bricks
10) Form cavity
11) Fill cavity
12) Earth filling
13) Damp-proof course.

Oversite

14) Level and compact
15) Hardcore
Blinding
Concrete bed
Tamp surface
Fabric reinforcement
16) Damp-proof membrane.

The take-off list is prepared after scrutinis the drawings and aims to identify what has be measured and the sequence of measurement.
A few minutes spent doing this will avoid many errors in practice.
As each item on the list is measured it car ticked off.

The following in work up to d.p.c. level

Topsoil strip

Foundation width	=	0.600
wall width	=	0.255
projection	=	0.345

Strip length

Wall length	4.000
add projection	0.345
	4.345

Strip width

Wall length	2.500
add projection	0.345
	2.845

External trench Length

2/4.345	8.690
2/2.845	5.690
	14.380

/.

Waste calculation to find:

1. Topsoil strip dimensions.
2. External trench side length.
3. Centre-line of trench (this is the same the cavity line in this case).
4. Internal trench side length.

Fig. 8.4.5

Dimensions	Squared	Description	Notes
		Centre Line of trench	
		External trench length 14.380	Deduct from external trench length 2/300 per corner.
		deduct 4/2/300 2.400	300 mm = half trench width.
		11.980	
		Internal trench length	
		External trench length 14.380	Deduct from external trench length 2/600 per corner.
		deduct 4/2/600 4.800	600 mm = trench width.
		9.580	
4.35 2.85	12.40	Excavate topsoil to be preserved average 150 mm deep.	*SMM D.9* Dimensions to extreme edge of foundation trench. Measured in m^2 stating average depth of excavation
4.35 2.85 0.15	1.86	Deposit preserved topsoil in temporary spoil heap for re-use average 20 m from excavation.	*SMM D.31* The same dimensions are used as for topsoil excavation but depth added as required in m^3.
11.98 0.60 0.75	5.39	Excavate foundation trench exceeding 0.30 m wide commencing at surface strip level maximum depth not exceeding 1.00 m	*SMM D.6.b, D.11* Centre line of trench × width of trench × depth of trench below surface strip level.
		&	
		Remove excavated material from site.	*SMM D.29*
14.38 0.90	12.94	Earthwork support maximum depth not exceeding 1.00 m and width between opposing faces not exceeding 2.00 m.	*SMM D.14, D.15, D17* Maximum depth given as clause D.1 and width between opposing faces as D.17. The depth of support to external trench is 900 mm deep whilst the internal trench is 750 mm deep.
9.58 0.75	7.19		
	20.13		
11.98 0.60	7.19	Level and compact base of excavation to receive concrete.	*SMM D.40* Measured in m^2. Centre-line of trench × width of trench.
11.98 0.60 0.75	5.39	Plain concrete in foundation trenches (1:3:6–19) over 300 mm thick poured against face of excavation.	*SMM F.6.2, F.5.2, F.4.6* Measured in m^3 stating thickness as clause F.5.2. Centre-line of trench × width of concrete × depth of concrete.

3.

TRENCH FILL FOUNDATION continued

Timesing	Dimensions	Squaring	Description	Notes
			Brickwork in foundations	
			Centre line of external skin	
			Perimeter of brickwork	Waste calculation to find centre-line of external skin of brickwork.
			2/4.000 8.000	
			2/2.500 5.000	
			13.000	
			deduct 4/2/51.25mm 0.410	
			12.950	
			Centre line of internal skin	Waste calculation to find centre-line length of internal skin.
			perimeter of brickwork 13.000	Note deduction = external wall = 102.5; cavity = 50; half internal skin = 51.25; 203.75
			deduct 4/2/203.75 mm 1.630	∴ deduct 2/203.75 per corner.
			11.370	Two methods of measuring the brickwork now described. In this the first the centre-lines of both the external and internal walls have been calculated. These have been timesed by the height of the common and facing bricks as appropriate. Many building students find the first method easier to understand whilst quantity surveyors would probably use the second method.
	11.37 0.30	3.41	Half brick thick wall in skins of hollow wall in calcium silicate bricks class 3 in stretcher bond in cement mortar (1:4)	*SMM G.5.1, G.5.3* Centre-line of internal skin × height of brickwork to d.p.c. level. Centre-line of external skin × height of one course of brickwork. The other 3 courses of the external skin will be measured in facing bricks.
	12.59 0.08	1.01		
		4.42		
	12.59 0.23	2.90	Half brick thick wall in skins of hollow walls entirely in facing bricks (P.C. £100 per 1000) in stretcher bond in cement mortar (1 : 4) pointed with a neat rubbed in joint.	*SMM G.14.9, G.5.3* Half brick and one brick thick walls entirely facing bricks are kept separate. Note description in skins of hollow walls. Centre-line of external skin × height of 3 courses of facing bricks. Height of one course of brickwork = 65 mm + 10 mm joint = 75 mm

4.

Timesing	Dimensions	Squaring	Description	Notes
			ALTERNATIVE METHOD (FOR BRICKWORK)	
2/	11.98 0.30	7.19	Half brick thick wall in skins of hollow walls in calcium silicate bricks class 3 in stretcher bond in cement mortar (1:4)	Here the centre-line of the cavity has been twice timesed. This gives the actual length of both skins. This can only be done where both skins have the same width dimensions.
	12.59 0.23	2.90	*Deduct* ditto & *Add* Half brick thick wall in skins of hollow walls entirely in facing bricks (P.C. £100 per 1000) in stretcher bond in cement mortar pointed with a neat rubbed in joint.	This adjustment is necessary to obtain the correct quantity of common and facing bricks. In the previous dimension the total brickwork was measured to d.p.c. level as if all in sand-limes. The adjustment corrects the quantities for commons and measures the area of facing bricks.
	11.98 0.30	3.59	Form 50 mm wide cavity in hollow wall including 4 number galvanised butterfly wall ties per m².	*SMM G.9* Given in m² stating width of cavity and type and spacing of wall ties in the description. Centre-line of cavity × height of cavity to d.p.c. level.
	11.98 0.05 0.15	0.09	Weak concrete (1 : 12) filling to hollow wall not exceeding 100 mm thick.	*SMM F.61, F.6.18, F.5.1* Concrete filling to cavity measured in m³ stating width as clause F.5.1. Cavity fill taken in this case to within 150 mm of d.p.c. level.
			External trench length = 14.380 Face of brickwork perimeter = 13.000 2)27.380 13.690	Waste calculation to find centre-line length of earth filling. External trench length and face brickwork perimeter are added together. The centre-line of earth fill is the mean of these two lengths.
			Width of earth fill foundation width = 600 wall width = 255 2) 345 projection = 172.5	Calculation to find width of earth filling.
	13.69 0.17 0.15	0.35	Earth filling around excavations with topsoil from temporary spoil heaps.	*SMM D.33, D.34* Measured in m³. Source of filling material to be classified as D.33.

5.

TRENCH FILL FOUNDATION continued

	Dimensions	Squared	Description	Notes
			<u>Oversite construction</u>	
			Strip length 4.350 deduct 2/600 1.200 3.150 Strip length 2.850 deduct 2/600 1.200 1.650	Waste calculation to find dimensions of leve and compact under hardcore bed. Trench width is deducted from strip length × 2 for each length.
	3.15 1.65	5.20	Level and compact surface of ground to receive hardcore.	*SMM D.40* Measured in m^2.
			Length of building = 4.000 deduct walls 2/255 = 0.510 3.490 Width of building = 2.500 deduct walls 2/255 = 0.510 1.990	Waste calculation to find internal sizes of building. Many of the oversite measurements are similar and can be anded on in the following manner.
	3.49 1.99	6.95	Hardcore filling in making up levels average 150 mm thick obtained off site	*SMM D.33, D.36* Over 250 mm thick measured in m^3. Under 250 mm thick measured in m^2 stating average thickness, and source of supply.
			&	
			Blind surface of hardcore with sand	*SMM D.43* Measured in m^2. Alternatively this could be included with filling where measured in m^2.
			&	
			Reinforced concrete bed (1 : 2 : 4 – 19) 100–150 mm thick laid on blinded hardcore bed (measured separately) (Cube × 0.15 = 1.04 m^3)	*SMM F.4.2, F.4.6, F.6.1, F.6.8, F.5.2* Note concrete bed measured in m^3. For convenience it has been anded on amongst other superficial dimensions. When squaring i carried out the cubic quantity is entered in t brackets.
			&	
			Tamp surface of unset concrete	*SMM F.9* Treating surface of unset concrete given in m stating type of treatment.
			&	
			Fabric reinforcement to BS 4483 weighing 3.05 kg/m^2 with minimum 150 mm end and side laps in ground slab (measured net no allowance for laps)	*SMM F.12.1, F.12.2, F.12.3, F.11.4* Measured in m^2. State weight and side and end laps. No allowance for cover to reinforcement has been made here.

6.

TRENCH FILL FOUNDATION continued

Timesing	Dimensions	Squaring	Description	Notes
			Damp-proof membrane	
			Internal length of building 3.490	Waste calculation to find sizes of damp-proof membrane. 50 mm has been added to the internal dimension of building to allow d.p.m. to turn under d.p.c.
			turn in under d.p.c. 2/50 0.100	
			3.590	
			Internal width of building 1.990	
			turn in under d.p.c. 2/50 0.100	
			2.090	
			Vertical girth	
			Brick perimeter 13.000	Waste calculation to find length of d.p.m. placed vertically.
			Deduct corners 4/2/255 2.040	
			10.960	
	3.59 2.09		1000 gauge polythene damp-proof membrane laid horizontally on blinded hard-core bed with minimum 200 mm laps (measured net no allowance for laps)	*SMM G.37.1, G.37.2* There is no specific mention in the SMM of damp-proof membranes. This description has been phrased using G.37 damp-proof courses. Note the quantity is measured net, the estimator must add for laps.
	10.96		Ditto but vertically to face of brickwork 150 mm deep.	*SMM G.37.1, G.37.2* The vertical part of the d.p.m. does not exceed 225 mm wide and is measured in linear metres.
2/	11.98	23.96	102.5 wide bituminous felt hessian damp-proof course to BS 743 in single horizontal layer with minimum 100 mm laps and bedded in cement mortar (1:4) (measured net no allowance for laps).	*SMM G.37* Damp-proof courses over 225 mm wide measured in m^2. Under 225 mm wide measured in linear metres stating width.

7.

Special Note: In order that the dimensions and descriptions of taking off can be speeded up abbreviations are used i.e. bwk = brickwork; ave = average and so on.

Also note that the squaring up has been done and that when descriptions have been incorporated into the Bill of Quantities they are cancelled out on the dimension paper.

Bill of quantities for Fig. 8.4.5

	Description	Qty	Unit	Rate	£	p
	EXCAVATION AND EARTHWORK					
	Preamble clauses would normally precede measured work.					
	Plant					
A	Allow for bringing to and removing from site all plant necessary for this section of the work.	item				
B	Allow for maintaining on site all plant required for this section of the work.	item				
	Site preparation					
C	Excavate topsoil to be preserved average 150 mm deep.	12	m^2			
	Excavation					
D	Excavate foundation trench exceeding 0.30 m wide commencing at surface strip level maximum depth not exceeding 1.00 m.	6	m^3			
	Earthwork support					
E	Earthwork support maximum depth not exceeding 1.00 m and width between opposing faces not exceeding 2.00 m.	20	m^2			
	Disposal of water					
F	Allow for keeping the surface of the site and the excavation free from surface water.	item				
	Disposal of excavated material					
G	Remove surplus excavated material from site.	6	m^3			
H	Deposit preserved topsoil in temporary spoil heap for re-use average 20 m from excavation.	2	m^3			
	Filling					
I	Earth filling around foundations with topsoil from permanent spoil heap.	1	m^3			
J	Hardcore filling obtained off site in making up levels average 150 mm thick.	7	m^2			
	Surface treatments					
K	Level and compact base of excavation to receive concrete.	7	m^2			
L	Level and compact surface of ground to receive hardcore filling.	5	m^2			
	Blinding					
M	Blind surface of hardcore with sand.	7	m^2			
	Protection					
N	Allow for protection to all work in this section.	item				
		To summary		£		

1.

CONCRETE WORK

Preamble clauses and plant items.

The following in OTHER CONCRETE work

approximate total volume 12 m^3.

	Description	Qty	Unit	£	p
	In-situ concrete				
A	Concrete (1 : 3 : 6–19) in foundation trenches over 300 mm thick poured against face of excavation.	5	m^3		
B	Reinforced concrete bed (1 : 2 : 4–19) 100–150 mm thick laid on blinded hardcore bed.	1	m^3		
C	Weak concrete (1 : 12) filling to hollow wall not exceeding 100 mm thick and under 1 m^3 in size.	1	nr.		
D	Tamp surface of unset concrete.	7	m^2		
	Protection				
E	Allow for protection to all work in this section.	item			
		To summary		£	

BRICKWORK AND BLOCKWORK

Preamble clauses and plant items.

	Description	Qty	Unit		£
	Foundations				
	Brickwork				
A	Half brick thick wall in skins of hollow walls in calcium silicate bricks class 3 in stretcher bond in cement mortar (1 : 4).	4	m^2		
B	Form 50 mm wide cavity in hollow wall including 4 number galvanised butterfly wall ties to BS 1243 per m^2.	4	m^2		
	Brick facework				
C	Half brick thick wall in skins of hollow wall entirely in facing bricks (Wealdon stocks) pc £200 per 1000 in stretcher bond in cement mortar (1 : 4) pointed with a neat rubbed-in joint.	3	m^2		
	Damp-proof courses				
D	102.5 mm wide bituminous felt hessian based damp-proof courses to BS 743 in single horizontal layer with minimum 100 mm laps and bedded in cement mortar (1 : 4) measured net, no allowance for laps.	24	m		
E	1000 gauge polythene damp-proof membrane laid horizontally on blinded hardcore bed with minimum 200 mm laps (measured net, no allowance for laps.)	8	m^2		
F	Ditto to vertical face of brickwork 150 mm deep.	11	m		
	Protection				
G	Allow for protection to all work in this section.	item			
		To summary			£

Conclusion

The remaining areas of work relating to the building, where excavation work, concreting work and brickwork have now been measured, would next be dimensioned to complete the total volume of work to be undertaken.

Remember that the Bill of Quantities is prepared to enable contractors to price the work precisely and using identical information.

Chapter 9

Contractor's pre-tender work

9.1 Types of contracts and standard forms

When a contracting firm receives an 'invitation to tender' an assessment of its workload and ability to undertake the type and size of the proposed project is essential before an application is made for the contract documents. If the firm decides that it is in an unfavourable position, it could then decline the invitation. Unfortunately, it has been known for a contractor to apply for the tender documents with the intention of submitting a tender which is excessive on the assumption that it would fail but at the same time keep his/her options open for future invitations from the architect.

The documents most usually received by a contractor on application are:

1. Drawings – sufficient to give the contractor an appreciation of the contract work.
2. Bills of Quantities – showing net work to be done..
3. Specifications – on materials and workmanship.
4. Tender Form – to be completed by the contractor showing the tender figure.

The contract drawings, copy of the appropriate JCT Standard Form of Building Contract and any other suitable particulars would be made available, usually at the architect's offices on a fixed date, for prospective tenderers to inspect to obtain a better appreciation of the magnitude of the contract. Contractors would also be given every opportunity to visit the site of the proposed works to note any relevant points such as location, accessibility, complexity, convenience, services availability, addresses of local public undertakers, public transport facilities, location of tips, local suppliers of materials, etc., local Job Centre for availability of labour, and any other point which may affect the estimator's estimate and ultimately the contractor's bid.

It must be appreciated that on being successful with a bid (tender) one of a number of Agreements can be made by the contractor with the client. An Agreement in the form of a contract is usually made with a professional architect or civil engineer presiding and who is acting for the client. Standard Forms are thereby used to bind the Agreement between the two parties, i.e. contractor and client.

Where a client prefers to deal directly with a contractor, particularly by agreeing to an 'all-in package deal', he/she assumes that it will lead to more favourable terms being offered by the contractor. The contractor who offers package deals normally employs his/her own design team which obviously contributes to better communications regarding the contractor's work; it would then be the contractor's responsibility to open suitable channels of communications with the client.

Clients normally approach two or more contractors initially to negotiate terms. The contractor when furnished with the client's brief would endeavour to produce a package deal which is more favourable than a competitor's to win the contract.

It is argued that the client is then at the mercy of the contractor especially when prices fluctuate and new legislation affects the contractor's position, thereby forcing him/her to take counter measures such as cutting corners and giving sub-standard work. While this may be true, the Agreement which is prepared by the contractor is so carefully worded as to safeguard both parties' positions, and in fact have been seen to work successfully for many numbers of years.

In construction work a contract is usually made either verbally or in writing. It is where two or more persons agree to act in a common cause and has legal implications. There must be an offer and an acceptance, which can be by one of the aforementioned methods.

During the course of many years businessmen have prepared Standard Forms either as an attempt to restrict the number of claims made upon them by dissatisfied customers, or have wanted to create a degree of goodwill leading to more business being offered in the future by satisfied customers. Fortunately, the latter case is in the majority, and particularly in the construction industry. Standard Forms have existed for many years to the mutual benefit of both parties. The Standard Forms are amended periodically as legislation and conditions change, thereby making improvements for both the seller and buyer.

The RIBA helped draft the first Standard Form of Contract assisted by the Building Employers Confederation (BEC), at the turn of the century and was then known, and continued to be so until the mid-1960s, as the RIBA Standard Form of Building Contract.

Other organisations later joined these bodies, such as: the Royal Institution of Chartered Surveyors, Association of Consulting Engineers, Association of County Councils, Association of Metropolitan Authorities, Association of District Councils, Greater London Council, Committee of Association of Specialist Engineering Contractors, Scottish Building Contract Committee and the Federation of Association of Specialists and Sub-Contractors. The committee representing their organisation became known as the Joint Contracts Tribunal, and the document prepared by this body is now called the JCT Standard Form of Building Contract.

A different form is used depending on the type of work, contract documents produced and client, and many Standard Forms have been prepared by other bodies such as that represented by the Federation of Civil Engineering Contractors, Institute of Civil Engineers and the Association of Consulting Engineers.

Examples of some of the Standard Forms of Contract are given below:

1. JCT Standard Form of Building Contract
   Private Edition with Quantities.
2. JCT Standard Form of Building Contract
   Private Edition without Quantities.
3. JCT Standard Form of Building Contract
   Private Edition with Approximate Quantities.
4. JCT Standard Form of Building Contract
   Local Authority Edition with Quantities.
5. JCT Standard Form of Building Contract
   Local Authority Edition without Quantities.
6. JCT Standard Form of Building Contract
   Local Authority Edition with Approximate Quantities.

There are supplement documents to be used with the aforementioned forms which are:

(*a*) Fluctuation Supplements clauses 38, 39 and 40 for Private Contracts.
(*b*) Fluctuation Supplements clauses 38, 39 and 40 for Local Authority Contracts.
(*c*) Formula Rules clause 40.
(*d*) Sectional Completion Supplement.
(*e*) Contractor's Designed Portion Supplement – where contractor is to design a portion of the works.

7. JCT Fixed fee Form of Prime Cost Contract.
8. JCT Agreement for Minor Works (and there is a Supplement).
9. JCT Agreement for Renovation Grant Works (when architect is or is not employed).
10. JCT Nominated Subcontractors/Suppliers Tenders, Warranties, Nominations and Subcontracts – documents NSC/1, NSC/2, NSC/2a, NSC/3, TNS/1, TNS/2, NSC/4, NSC/4a including the Fluctuations and Formula Rules Supplements.
11. JCT Standard Form with Contractor's Design (and the Formula Rules Supplement).

12. BEC Nominated 'Green' Subcontract (if JCT NSC Form is not used).
13. BEC Domestic 'Blue' Subcontract (with articles, conditions, Formula Rules, etc or use DOM1 or DOM2 Forms).
14. GC/WKS/1 Standard Form (General Conditions for Government Contracts for Building and Civil Engineering).
15. GC/WKS/2 Standard Form (for minor Government works).
16. ICE Conditions of Contract (the Institution of Civil Engineers Form).
17. Management Contracting Agreement.

The various Standard Forms of Contract contain clauses which have been inserted to safeguard the interests of both parties to the contract, examples of which are outlined under Conditions of Contract numbered 1 to 37(which follow on pages 230 to 237), with Clauses 38, 39 and 40 contained in supplement documents known as Fluctuations and Formula Rules Supplements.

The JCT Standard Form of Building Contract (Private Edition with Quantities)

When Bills of Quantities are included with other contract documents for proposed building projects and which are to be executed for clients other than local or central Governments, then this document is suitable and is divided into nine main sections (Parts 2–6 containing the clauses dealt with at the end of the list):

1. *Articles of Agreement:* State when the Agreement was made between the employer and contractor, including their names and addresses.

- *Recitals.* First, allows for details on the type of works/building and address, and whose direction the Drawings and Bills of Quantities were prepared. Second, confirms that the contractor has supplied the employer with a priced Bill of Quantities. Third, shows the number of drawings provided called the Contract Drawings. Fourth, gives status of employer for tax purposes.
- *Articles.* 1. Contractor's obligations.
  2. Contract sum.
  3. The architect.
  4. The quantity surveyor.
  5. Settlement of disputes – Arbitration.
- *Signature spaces* for employer's and contractor's representatives' signature and of witnesses' signatures.

2. *Conditions: Part 1 – General (Clauses 1–34).*

3. *Conditions: Part 2 – Nominated subcontractors and nominated suppliers (Clauses 35–36).*

4. *Conditions: Part 3 – Fluctuations (Clauses 37–38).*

5. *Labour and material cost and tax fluctuations (Clause 39).*

6. *Use of Price Adjustment Formula (Clause 40).*
7. *Appendix:* contains additional information (to be written in) such as: statutory tax deduction scheme, arbitration, date of completion, defects liability period (usually six months), insurance cover for any single event, professional fees percentage cover, date of possession, liquidated and ascertained damages, period of delay, period of interim certificates, retention percentage (if less than 5 per cent), period of valuation for final certificate, work reserved for nominated subcontractor for which the contractor desires to tender, fluctuations, percentage addition, formula rules.
8. *Supplemental provisions* (the VAT Agreement): This section lays down rules which the contractor and employer should observe relating to all aspects of VAT inclusions in payments made for services rendered.

Contractors must be familiar with the JCT Forms, and in addition should be able to interpret the clauses. It must be understood that if one or more of the clauses is unsuitable or inappropriate for a proposed contract it may be deleted by the architect, with the client's approval; the contractor should therefore check for deletions or additions or inclusions before signing.

Conditions of Contract (JCT Form), in brief

Conditions: Part 1 – General.

Clause 1: Interpretations

Meanings are given to certain words used within the various clauses throughout the conditions sections, etc.

Clause 2: Contractor's obligations

This clause states that the works shown on the drawings and described in the Bills of Quantities are to be completed to the reasonable satisfaction of the architect. Any discrepancies found between the drawings and bills must be notified in writing to the architect who shall then issue instructions.

Clause 3: Contract sum – additions or deductions – adjustments – interim certificates

When an adjustment of the contract sum is allowed for during the construction work it should be taken into account within the next interim certificate.

Clause 4: Architect's instructions

The contractor is to comply with all instructions issued by the architect which are relevant to the works. The instructions should be in writing, and if made verbally shall be confirmed in writing by the contractor within seven days (if the architect fails to confirm them himself/herself), whereon the architect should acknowledge the confirmation within a further seven days.

Inspection copies of the documents should be retained by the architect for the client and contractor's inspection.

Although it states that drawings and other documents should be returned to the architect, if so requested by him/her, at the completion of the project, it also states that use should not be made of the documents by the contractor other than for executing the works. (It is important to note that the Copyright Act 1956 (and as amended in 1982) would vest the drawings with the architect.) The contract documents to be issued to the contractor before work commences on site are:

1. Articles of Agreement and Conditions of Building Contract (one copy).
2. Contract drawings (two copies).
3. Contract Bills of Quantities — unpriced (two copies).
4. Descriptive schedules or other relevant documents for use in carrying out the works (two copies of each).

The contractor is charged with providing the architect with two copies of a master programme for the works. Also the contractor should keep one copy of all the aforementioned documents on-site for reference by the architect during his visits.

Clause 6: Statutory obligation, notices, fees and charges

Under the various Acts of Parliament, Statutes are introduced which may place obligations on the building owner or contractor during construction work and must be complied with. Byelaws and Regulations are also in existence and are enforced by either local authorities or statutory authorities such as gas, water and electricity undertakings.

Note: Below are some examples where notices and fees are involved.

1. Rates for site huts.
2. Licences for hoardings.
3. Crossovers licences.
4. Notices required to the building control offices when work commences and at inspection stages.
5. Applications to statutory authorities for temporary services, etc.

Notification must be given to the architect by the contractor if, in complying with the Regulations, Rules, Orders, etc. alterations are necessary to the drawings or bills.

Clause 7: Levels and setting out of the works

The architect is responsible for establishing levels for the works and the contractor is responsible for the setting-out from the levels.

 Clause 8: Materials, goods and workmanship to conform with description, testing and inspection

Materials used on-site should conform to the standards in the contract bills, as should goods and workmanship. If the architect in carrying out inspections or tests requires the contractor to open up any works, say, to inspect part of the foundations, the extra costs incurred would be added to the contract sum unless faulty work was discovered, then the contractor would be liable for rectification costs.

Clause 9: Royalties and patent rights

Contractors are not normally liable for claims made against them if they are instructed by the architect to use a patented article. Any legal claims made against them would be charged to the contract, but the contractors would be liable if they have taken it upon themselves to use an article which is later discovered to be patented.

Clause 10: Person-in-charge

A competent person who is able to act for the contractor and who understands instructions from the architect should be available on the works.

Clause 11: Access for architect to the works

Reasonable access must be afforded for the architect to inspect the works which also includes workshops of the main contractor and of the sub-contractors where components are being prepared.

Clause 12: Clerk of works

The clerk of works acts solely as an inspector on behalf of the client under the directions of the architect. Instructions given by the clerk of works to the person-in-charge are not operative until they are confirmed in writing by the architect within two days.

Clause 13: Variations and provisional sums

The architect is entitled to make variations to the building design during the construction stage, and the contractor can also make variations with his approval.

A variation is a change in the design, quality or quantity of the works, and where the contractor incurs additional expenses he/she will be able to recoup them from the client.

Where provisional sums are included in the Contract Bills, the architect will instruct the contractor at the appropriate time on this form of expenditure. Where a variation of work cannot be accurately measured, daywork rates would be allowed to the contractor as previously agreed, or if no agreement were made, it would be calculated using the Definition of Prime Cost of Dayworks carried out under a Building Contract, with percentage additions as outlined in the Appendices of the JCT Standard Form of Contract.

Clause 14: Contract sum

After the signing of the contract, if errors are found in the calculation of the contract sum, no alteration shall be made and the contract shall still be binding.

Clause 15: Value added tax – supplemental provisions

States that references made to the contract sum shall be exclusive of VAT. The employer will therefore be liable for payment for VAT to the contractor on work which is rated.

Clause 16: Materials and goods unfixed or off-site

Materials shall not be removed from site unless the architect gives his approval in writing, and if an interim certificate has been issued which takes into account any materials, then they should only be used exclusively for the client's work. In both cases the contractor is still liable for the loss or damage of the materials.

Clause 17: Practical completion and defects liability

When the architect is satisfied that the work on-site is completed, he will issue a certificate and the employer would then take possession of the works. At the expiration of the defects liability period (usually six months after the practical completion date), a schedule of defects should be prepared by the architect within fourteen days for presentation to the contractor who would then be expected to make good the faults. When the faults have been made good a certificate (Certificate of Completion of Making Good Defects) shall then be issued by the architect to the contractor.

Clause 18: Partial possession by employer

Where it is agreed, the works can be handed over to the client in stages. Each stage so completed would be valued and the date on which the valuation certificate is issued would be the start of the defects liability period. This means that there could be numerous practical completion dates and defects liability periods each of which would be subject to the points outlined in Clause 17. The client would also become liable for the insurance of the part of the building of which he/she takes possession.

Clause 19: Assignment and subcontracts

To assign the contract requires the written consent of both parties to the contract and where the contractor wishes to subcontract some of the work, written permission from the architect is required.

Clause 20: Injury to persons and property and employer's indemnity

If any claims are made against the employer arising from loss or injuries sustained by anyone and which is caused by the contractor or his employees while carrying out the works, the contractor is bound to indemnify the employer unless the injuries, etc. were caused by the employer's negligence.

Clause 21: Insurance against injury to persons and property

It is the responsibility of the contractor to take out adequate insurance cover against the possibility of claims being made as in Clause 20, and the contractor shall provide documentary evidence to the architect to this effect.

Clause 22: Insurance of the works against perils

Either the contractor or the employer is responsible for the insurance of the works and materials against damage by fire, explosions, storm, riot, etc. the responsibility of which would have to be agreed by both parties. The contractor's equipment is not considered as part of the works.

Under Clause 22C, insurance cover for alteration work, etc. to existing buildings is the responsibility of the employer.

Insurances should be taken out to operate until Practical Completion.

Clause 23: Date of possession completion and postponement

The contractor should be given possession of the site and should complete the works on or by the dates agreed and as shown in the Appendix of the JCT Standard Form of Building Contract. This is subject to the Extension of Time clause.

The architect is empowered to postpone any work.

Clause 24: Damages for non-completion

The contractor may have to pay Liquidated and Ascertained Damages for failure to complete the contract on time. The damages are usually calculated taking into consideration the employer's loss of profits through delay in opening up, say, his new shop, and any resulting inconvenience.

Clause 25: Extension of time

Written notice should be sent by the contractor to the architect where there is a delay to the works due to conditions outside the contractor's control, and where the architect foresees that the contract duration needs to be extended a written extension of time will be allowed.

The conditions which normally allows the contractor to prepare written notices are: unusually poor weather; delay in sending instructions by the architect; delays in the execution of work by nominated subcontractors, suppliers and artists; industrial action by operatives or others associated with the works.

Clause 26: Loss and expense caused by matters materially affecting regular progress of the works

Contractors can claim in writing for any loss or expenses incurred due to disturbances in the progress of the work such as: opening up the works for inspection by the architect which proves to be satisfactory; for delays in receiving instructions from the architect; and delays caused through nominated subcontractors, etc.

Clause 27: Determination by employer

When the contractor fails to carry on normally with the works (working regularly and in good faith) or becomes a bankrupt, the employer may cancel the contract and employ another firm to continue the work thereby charging the contractor for any additional expenses. If there is a balance in favour of the contractor after deducting expenses from the Retention held by the employer, it would be paid to the contractor.

Clause 28: Determination by contractor

When the employer creates problems resulting in the delay of payment on any certificates or the value of a certificate is not paid within fourteen days, the contractor can determine the contract seven days after sending notice that he/she intends to determine the contract. There are other reasons whereby the contract can be determined, these are: on the bankruptcy of the employer; when work is suspended due to delays by nominated subcontractors/suppliers; civil commotions, etc.

The contractor will have the right to claim for that part of the works which has been executed, and for expenses in determining the contract.

Clause 29: Works by employer or persons employed or engaged by employer

If certain work has to be done which is outside the work of the contract and the contract bills by the employer (client) or persons employed by him/her, the contractor should permit such work. Persons employed by the client are the sole responsibility of the client.

Clause 30: Certificates and payments

The architect shall issue interim certificates at intervals (usually monthly) as agreed and shown in the Appendix of the JCT Standard Form of Building Contract and certificates should be honoured within fourteen days by the employer. The amount to be paid on each certificate should be the total value of work done and value of certain materials off- and on-site less any previous payments and retention. The retention (kept by the employer) should be not more than 5 per cent of the contract sum.

A certificate is issued at the practical completion stage to verify the commencement of the defect liability period. Another certificate is then issued by the architect which releases half of the total retention to the contractor, and the employer would again have to honour it within fourteen days.

The final certificate would be issued by the architect to the contractor not earlier than three months after the defects liability period or following the making good of all defects by the contractor, whichever is the later. This certificate is evidence of the works being completed satisfactorily.

Clause 31: Finance (No. 2) Act 1975 – Statutory tax deduction scheme

The clause makes reference to the Income Tax (Subcontractors in the Construction Industry) Regulations 1975. It outlines when the contractor should or should not deduct tax from a subcontractor before paying money for work done under a contract.

Clause 32: Outbreaks of hostilities

Where there is a general mobilisation of HM armed forces, either side to the contract can determine the contract.

Clause 33: War damage

Where the contract is not determined upon the outbreak of hostilities any damage so resulting may be repaired by the contractor on instructions being given by the architect, and it would be classed as a variation.

Clause 34: Antiquities

On the discovery of objects of interest or historical significance on-site the contractor should preserve them until instructions are issued by the architect. Expenses incurred by the contractor in doing so can be claimed from the employer. The objects, etc. become the property of the employer.

Part 2: Nominated subcontractors and nominated suppliers

Clause 35: Nominated subcontractors

A nominated subcontractor is one which is appointed by the architect or client and who is acceptable to the contractor. In the Bills of Quantities an architect could make provisions in the form of a prime cost sum or provisional sum where normally it is intended that a known subcontractor is to undertake work of a specialist nature.

Nominated subcontractors enter into a contract using a nominated subcontract form (NSC 1, 2, 3 or 4) which binds them to agree to: do work to the satisfaction of the contractor and architect; insure themselves against claims; complete their work on time; indemnify the main contractor against substandard work and allow 2½ per cent cash discount to the main contractor on each certificate paid.

Clause 36: Nominated suppliers

These suppliers are the architect's/client's choice for the supply of materials and equipment as outlined in the Bills of Quantities under Prime Cost Sums, or where the architect instructs on expenditure under Provisional Sums, and 5 per cent cash discount should be allowed to the contractor on payments made by the contractor to the supplier usually within thirty days from the receipt of an invoice/statement.

Part 3: Fluctuation

Clause 37

The fluctuations are dealt with in accordance with whichever of the following alternates are to apply:

Clause 38

Increases or decreases to levies, tax, insurances, caused by Government legislation, should be allowed to or by the contractor if they occur after the tender date.

Clause 39

In addition to increases or decreases as outlined in Clause 38, increases or decreases in the rates of wages and materials will be allowed to or by the contractor if they occur after the date of tender.

Clause 40

Fluctuations to taxes, insurances, wages and materials after the date of tender would be allowed but by using a system known as the Formula Rules/Indices. It is an alternative system to Clause 39, but still allows full fluctuations to be paid to or by the contractor.

The Supplements (separate documents) deal with the Clauses 38, 39 and 40.

If the parties to the contract wish that Clause 38 should apply, then Clauses 39 and 40 should be deleted. Similar action should be taken if either Clause 39 or 40 is to apply.

9.2 Bills of Quantities

(See also Section 8.4 on page 211.)

Tender preparation

On receipt of the Bills of Quantities, including the other tender documents, the Estimator's work is made easier if other members of the contractor's team, i.e. contracts manager, plant manager and planner, are invited to agree on the following:

1. Sequence and method of constructing the building.
2. The proposed plant to be used, and whether it is to be the contractor's own or hired from outside the business.
3. The number of key supervisors required.
4. The volume of work to be sub-let (subcontracted out).
5. The parts of the Site Survey Report which will affect prices (a visit to the site by a surveyor is important to establish general site conditions and to note any relevant matters).
6. A programme of work (Pre-Tender Programme).

A detailed study of the Bills will reveal many details about the proposed contract which will assist the Estimator to prepare an estimate, such as:

1. The Scope of the Works – a general appreciation of what type of building is involved.

2. Conditions of Contract – which JCT Schedules of Conditions apply? But in particular, which fluctuation clause is to operate? If no fluctuations are allowed, prices may have to be adjusted to take into consideration inflation.
3. Duration of the Contract – may require the employment of complete new gangs of operatives if contract period is excessive.
4. Percentage Retentions – affects cash flow.
5. Liquidated and Ascertained Damages – if the figure is high, then costs may have to be adjusted to allow for bonus payments and overtime to ensure contract is completed on time.
6. Value of Nominated Sub-Contractors' Work.
7. Value of goods to be supplied by Nominated Suppliers.
8. Major Sections of the work to be undertaken by the main contractor.

With the aforementioned information in mind the Estimator will be in a position to make a start on the Estimate for the work to be done. This of course depends on the quotations received from those subcontractors and suppliers who will be invited to tender for work, etc. which is to be sub-let by the main contractor.

Enquiries for quotations

A schedule of materials is prepared by the Contractor's Estimator to enable enquiries for quotations to be sent to a select list of suppliers – the lowest bid being accepted. Similarly, a list of subcontracting work should be made and quotations would be invited from specialist subcontractors. Photocopies of the appropriate sections of the Bills and Specifications, say, the whole of the Electrical Installations Section for the list of electrical subcontracting firms, must be included along with a questionnaire and copy of the Pre-Tender Programme, to enable the subcontractors to give fair quotations and assess their work-load and make possible adjustments to fit in with the main contractor's programme.

Generally, the lowest quotation/tender received by the main contractor will be accepted having taken into consideration other factors. A low quotation from a supplier however may mean that it is expected of the main contractor to pay for delivery as extra, while a slightly higher quotation includes delivery and unloading on-site and is on the whole cheaper: also, discounts offered by the supplier or subcontractor affects the overall quotation.

Prime Costs and Provisional Sums which are already included in the Bills for work to be executed and for materials to be supplied by the architect's/client's nominated subcontractors or suppliers will be added to the rates worked out by the main contractor's Estimator. The total amount is then known as the Estimate, i.e. the amount it will cost the contractor to do the work without the additions of overheads and profit.

The next stage would be to decide on percentage additions for overheads and profit which is added to the Estimate to give the final Tender Figure. These additions would depend on the cost of supporting a head office and how keen the firm is to win the contract bearing in mind the strength of its competitors and its success rate in the past for contracts of a similar nature.

Bills of Quantities

(See also Section 8.4)

Most Bills are divided into numerous sections, each of which are clearly laid out depending on which Standard Method of Measurement was used in its preparation. The various SMM are:

1. Standard Method of Measurement of Building Works.
2. Standard Method of Measurement of Civil Engineering Works.
3. Department of the Environment Method of Measurement for Road and Bridge Works.
4. The Code for the Measurement of Building Works in Small Dwellings.

The SMM of Building Works is used perhaps more than any other and is adopted to prepare Bills of Quantities for building work only. It is a document which clearly lays down rules by which the professional or other quantity surveyors should follow when preparing Bills. The general layout of Bills of Quantities follows the pattern suggested in the SMM of Building Works which states that the Bills (showing all the materials measured net, as fixed in position or removed, disregarding any labour or wastage) should be presented in the following order:

1. Preliminaries – sometimes followed by the Specifications.
2. Demolition.
3. Excavation and Earthworks.
4. Piling and Diaphragm Walling.
5. Concrete Work.
6. Brickwork and Blockwork.
7. Underpinning.
8. Rubble Walling.
9. Masonry.
10. Asphalt Work.
11. Roofing.
12. Woodwork.
13. Structural Steelwork.
14. Metalwork.
15. Plumbing and Mechanical Engineering Installations.
16. Electrical Installations.
17. Floor, Wall and Ceiling Finishes.
18. Glazing.
19. Painting and Decorating.
20. Drainage.
21. Fencing.

Each section, as outlined in the SMM, is presented as a separate Bill in Bills of Quantities produced by professional Quantity Surveyors where large works are planned. On smaller contracts two or more sections may be grouped together under one Bill. Bills of Quantities follow, however, the general pattern as follows:

240 *Bill No. 1 – Preliminaries:* This shows, as an introduction to the Bills of Quantities, the items laid down in the SMM of Building Works, which are:

Preliminary particulars

(*a*) Name and addresses of the employer, architect and quantity surveyor.
(*b*) Description of site – means of access; where keys can be obtained; where drawings, etc. can be inspected.
(*c*) Description of works – what type of building is proposed, including the plan sizes and height.

Contract particulars

(*a*) Conditions of Contract – the type of contract should be outlined stating which Standard Form will apply.
(*b*) Joint Contract Tribunal Schedule of Conditions – they must be given, such as those listed in Section 9.1 from Clauses 1 to 40.
(*c*) The Appendix to the Conditions of the contract – they are normally inserted relating to the following:
   i Dayworks – Percentage Additions on Prime Costs – where the architect instructs the contractor on daywork expenditure then the contractor would be entitled to add a percentage for overheads and profit to the labour, materials and plant used to carry out the work.
   ii Defects Liability Period – usually six months allowed.
   iii Date of Possession of Site – the date to enable the contractor to start work.
   iv Date of Completion – usually states, say, 'within fifty weeks from date of possession'.
   v Liquidated and Ascertained Damages – an amount is inserted which the contractor is expected to pay for every week that work on the contract extends beyond the agreed date of completion.
   vi Period of Delay – a maximum of three months is allowed for delays due to damages caused to the Works by fire as outlined in the JCT form, Clause 25. A delay of up to a maximum of one month would be allowed in any other case.
   vii Prime Cost Sums for which the contractor wishes to tender – contractors are given the opportunity to tender for some work which would normally be offered to a Nominated Subcontractor.
   viii Period of Interim Certificates – usually issued monthly.
   ix Period for Honouring of Certificates – client normally has to pay within fourteen days of interim certificate being issued.
   x Percentage of Retention – a maximum of 5 per cent on each payment made to the contractor until the work is completed to the satisfaction of the architect would be the value retained by the client.
   xi Limit of Retention – a percentage, say 2½ per cent of the contract sum, would be the maximum amount which the clients would be allowed to retain until the end of the defects liability period.
   xii Percentage Additions – the amount which the contractor would be entitled to on fluctuations, if these were allowed.

General matters: Items, for convenience in pricing, should be given at this stage in Bill No. 1, such as: plant; safety, health and welfare details; arrangement for notices and fees to public authorities; setting out the works; security of site; complying with police regulations; water for the works; contractor's and employer's liability insurances.

Temporary works: Temporary Works are included at this stage, such as: provision of temporary roads; sheds; messrooms; foreman's and clerk of works' offices and associated equipment; telephones; hoardings; scaffolding for the works; and other requirements such as protecting and cleaning the works.

Item	Bill No. 12					
	PRIME COSTS AND PROVISIONAL SUMS					
	Prime costs sums – subcontractors					
A	Provide the P.C. Sum of £1000 for steelwork to be provided and erected by a subcontractor to be nominated by the Architect.				1000	00
B	Allow for profit.	Item				
C	Allow for general attendance.	Item				
D	Allow for special attendance.	Item				
	Prime cost sums – suppliers					
E	Provide the P.C. Sum of £250 for fibreglass domelights to be delivered to site by a supplier nominated by the Architect.				250	00
F	Allow for profit.	Item				
	Provisional sums					
G	Provide the following Provisional Sums to be expended as directed by the Architect, the whole or part to be deducted if not required.					
H	Contingencies – Five thousand pounds.				5000	00
J	Stone laying ceremony – Three hundred pounds.				300	00
	Student note: *General attendance* – means allowing the subcontractor the use of scaffolding, messrooms, welfare facilities and other existing site facilities. *Special attendance* – means unloading and placing materials into position, and providing power and any other special requirements. *Contingencies*– for unforeseen work.					
	Carried to collection			£	6550	00

Fig. 9.2.1

Further provisions may be insisted on either at the contractor's own expense or the client's, such as: nameboards; bar charts to show programme of work; and reinstatement of surrounding adjoining areas.

Bill No. 2 – Specifications: As dealt with in Section 9.3.

Bill No. 3 – Demolition: If included in the contract work.

Bill No. 4 – Excavations and earthworks, and Bills No. 5, 6, etc. in the order as laid down in the Standard Method of Measurement.

A customary practice where various PC Sums and Provisional Sums have been included in the Bills would be to summarise them on the last pages in the Bills of Quantities to highlight to the contractor the extent of the work which is to be executed by Nominated Subcontractors, and where special materials are to be supplied by Nominated Suppliers these would also be incorporated. An alternative to the previously mentioned method would be to include a separate Bill – usually at rear of the Bills of Quantities – to include all PC Sums, etc. which would show the work to be done by a subcontractor and space would then be made available for the main contractor to add a percentage (usually 2½ per cent) for profit, and further sums for general and special attendances. Where Nominated Supplies are involved, 5 per cent may be added for profit (see Fig. 9.2.1).

9.3 Specifications

When a contracting firm has received the proposed contract particulars, specifications, etc. in order to tender for work its representatives may discover one of three things:

1. The Specifications received are contained in a separate document to the Bills of Quantities.
2. The Specifications are incorporated within the Bills of Quantities either as a
   (*a*) Bill itself, or
   (*b*) Preamble to each section in the Bills of Quantities.
3. There is a Specifications document but Bills of Quantities are not included.

Specifications are sometimes referred to as Preambles to Trades; General Descriptions; or Materials and Workmanship, and may be written in a number of ways depending on the current practice of the architectural firm. Each section of work, as laid down in the Standard Method of Measurement of Building Works, is dealt with under one of the following methods:

1. The Specifications are divided into trades and each trade is dealt with separately under Materials and Workmanship, e.g.
   (*a*) Concrete Work: i. materials; ii. workmanship.
   (*b*) Brickwork and Blockwork: i. materials; ii. workmanship.
   (*c*) Roofing: i. materials; ii. workmanship.

(See Figs. 9.3.1 to 9.3.4 for details of the brickwork and blockwork section, using this system.)

2. The Specifications are divided into two parts, the earlier part shows the 'Materials Section', giving the exact descriptions of all materials to be used, quoting relevant British Standards; and the later part shows the 'Workmanship Section', describing how the materials should be protected, fixed, and precautions to be taken quoting any BS Codes of Practice.
3. The Specifications are divided into trades, and materials and workmanship are combined under each trade without using separate sub-headings of Materials and Workmanship.

Item	Specifications				£	p
	BRICKWORK AND BLOCKWORK					
	Materials					
A	*Foundation Bricks* Bricks in foundations shall be Warnham Selected hard pressed clay bricks with a crushing strength of not less than 32 N/mm².					
B	*Common Bricks* When used in superstructure they shall be Flettons or other approved bricks to comply with BS 3921.					
C	*Facing Bricks* The Redland Heather Grey Holbrook perforated facing bricks rustic texture as manufactured by Redland Bricks Ltd. are to be used.					
D	All bricks shall be good clay, hard, square, well burnt, free from all defects, uniform in size and shall comply with BS 3921 unless otherwise described. They shall be obtained from an approved manufacturer and samples are to be submitted to the architect for approval.					
E	*Blocks* Internal skins of external walls and partition walls shall be 'Thermalite-Ytong' blocks obtainable from Thermalite Ytong Ltd and shall be to BS 2028 type B.					
F	*Cement* All cement used in the works shall be Ordinary or Rapid Hardening Portland Cement conforming to BS 12 requirements. The cement shall be delivered in 50 kg sealed bags and marked with the manufacturers name. Deliveries in bulk shall not be made without prior consent from the architect. Cement should be stored in a hut which is raised from the ground and weatherproofed.					
	Carried to Collection			£		
	(Bill & page no.) 2/6					

Fig. 9.3.1

This latter method appears to be used widely in Building Works, while method number two is popular in civil engineering works. The first method is a compromise between numbers two and three.

Checking the specifications by contractor

On receipt of the contract Specifications from the architect, certain points should be noted before the contractor prices the bills of quantities and prepares the tender figure, such as:

Item	Specifications				£	p
	Brickwork, etc. (Contd.)					
A	*Lime* To be hydrated lime conforming to BS 890 class B, and shall be delivered in a fresh condition in sealed bags bearing the name of the manufacturer and shall be soaked in water not less than 16 hours before use.					
B	*Sand* This shall be clean, sharp, siliceous sand free from shell, limestone, clay, dirt or any other harmful impurity or material in sufficient quantity to affect adversely the durability and strength of the mortar, and shall comply with the BS 1200 table 1.					
C	*Cement Mortar* This shall comprise of one part cement and three parts of sand.					
D	*Composition Mortar* This shall comprise of one part cement, one part lime putty and six parts of sand.					
E	*Ties* Copper butterfly wire wall ties to comply with BS 1243, type B, and 150 mm long for 50 mm cavity.					
F	*Damp Proof Course* Shall be bitumen as Bs 743, type 5A, hessian base and joints well lapped.					
G	**Workmanship** All bricks and blocks to be neatly stacked on site and not tipped and should be protected from the elements.					
H	All mortar to be mixed on site in mechanical mixers adding only sufficient water to give correct consistency.					
	Carried to Collection			£		
	2/7					

Fig. 9.3.2

1. Are the Specifications very stringent and what experience has the firm had with the architect on previous contracts?
2. To ensure that the specifications are compatible with the items given in the bills of quantities.
3. Where BSS and BSCP are quoted, check on their availability in the office to verify exact requirements, particularly if details given are vague.
4. What special or out of the ordinary specification requirements are there?
5. The names of any specialist firms which have been 'Nominated' to undertake stages or sections of the Works so that enquiries for subcontracting tenders can be sent to them.

Item	Specifications				£	p
	Brickwork, etc. (Contd.)					
A	No mortar which has commenced to set shall be used or remixed, but shall be rejected.					
B	Additives not allowed in mortar unless approval first given by the architect.					
C	Bricks and blocks to be wetted as necessary during hot or dry weather.					
D	Brickwork shall be built in English bond or Stretcher bond for half skin walls unless otherwise described.					
E	Brickwork to be built frogs upwards and filled with mortar and perpends or bed joints not to exceed 10 mm and built solid with mortar.					
F	Facing work to be kept clean and clear of mortar as work proceeds.					
G	Joints of facings to be pointed as work proceeds and should be flushed up and finished with key joint.					
H	*Hollow Walls* Cavity to be kept free of rubbish and mortar droppings by moveable cavity laths. Openings are to be left at base of each run of cavity to permit cleaning out at completion, and holes filled up afterwards to match existing brickwork. Wall-ties to be laid at 900 mm intervals horizontally and 450 mm vertically and at sides of openings of unbonded jambs vertical intervals to be not more than 300 mm. Ties to lay falling outwards to external skin of wall.					
I	No four courses of brickwork, when laid, to rise by more than 300 mm.					
	Carried to Collection			£		
	2/8					

Fig. 9.3.3

6. Observance of the specialist suppliers of materials and components named in the Specifications, whether as Nominated Suppliers or just suggested suppliers.
7. What special plant is mentioned which may not be shown clearly in the 'Bills'?
8. Checking any descriptions which state that the Standard Method of Measurement of Building Works has not been followed regarding a section in a 'Bill'.
9. To note the section dealing with the supply of materials which require 'written guarantees' from the contractor's own suppliers or sub-contractors.
10. Checking for any 'testing' or 'inspection' requirements by the architect which will affect the prices.
11. Paying due regard to sections requiring the furnishing of 'samples' to the architect.
12. Inspecting for 'special treatment' requirements, i.e. preservatives to soft-wood, windows, doors and structural timbers.
13. Tolerances expected on the structure, particularly regarding industrial-ised buildings, etc.

A well-written Specification leaves nothing to chance and the contractor then knows exactly what is expected and can price accordingly.

Item	Specifications **Brickwork, etc. (Contd.)**				£	p
A	Each stage of brickwork not to rise more than 1.5 m and racking back to be provided for continuation work.					
B	No work to be carried out in frosty weather unless permission is first obtained from the architect.					
C	Work should be covered up to protect it against the elements at the end of the working day or when weather necessitates.					
	Carried to Collection			£		
	2/9					

Fig. 9.3.4

9.4 Tender programme

It is sometimes a stipulation that an Outline or Tender Programme in the form of a bar chart is submitted to the architect at the same time as the Tender Figure and priced Bills of Quantities. The Programme should naturally agree with the contract period shown in the Appendix particulars to the JCT Form and contained in the Preliminaries section of the Bills of Quantities.

Although the lowest Tender Figure is normally accepted, the architect reserves the right to make enquiries about the financial stability and organisational ability of the successful tenderer.

To prevent unstable firms from applying for contract work which they may never be able to complete thereby incurring unnecessary expenses to the client in the event of the contractor's 'determination of the contract', a 'Performance or Guarantee Bond' (usually 10 per cent of the tender figure) is sometimes necessary which requires the contractor to find a bank or financier willing to act as guarantor in the event of the contractor being unable to meet his/her commitments to the client. If a firm were unstable financially, it would find difficulty in arranging a Bond and the next lowest Tender would then be accepted.

The contents of the Tender Programme is also important to show particularly the sequence of the work, and in its preparation it would assist the architect to form a further opinion of the contractor's organisation ability.

When preparing a Tender Programme it is important for the programmer/planner to form an appreciation of the proposed works by first studying the Drawings, Bills and the JCT Form. In discussing the project with other members of the team, it is essential to draft out particulars which will also be used later – in the event of the Tender being successful – such as:

1. Brief Method Statement.
2. Outline Plant Schedule – for major items of work.
3. Schedule of Accommodation – huts, etc.
4. Site Organisation Structure – personnel.
5. Subcontractors' Work.

These factors will be reappraised later if and when the Master/Contract Programme is prepared, whether in the form of a Bar Chart, Network Chart (Critical Path), Precedent Diagram or Line of Balance Chart – the use of which depends on the size and type of works involved.

The Tender Programme durations (length of time to do operations) for each stage/operation of work can be calculated scientifically, but because of the time and costs involved approximate values can be used instead, bearing in mind past experiences on similar work. When there is doubt about a duration, it would be prudent to check by calculations, e.g. Substructure Work.

Assuming that there is no doubt about the durations for excavations, brickwork to dpc, and concrete floor which amounts to, say, 9½ weeks and which agrees with a similar previous contract under similar conditions, the foundations this time are to be made using ready-mix concrete.

Measuring from the Drawings or extracting quantities from the Bills, the foundations are found to require 180 m^3 of concrete. The Programming Officer would now have to consider the following in calculating durations:

1. Type of plant, if required, to transport concrete where ready-mix trucks will find the foundations inaccessible.

Tender programme

Contract: Two Storey Office Block.

Contractor: G. Clare Ltd., The Croft, Cambs.

Client: B. Helen.

Architect: D. Joanne. A.R.I.B.A.

Date: 15th June 19

By: S. Marrs.

Op. No.	Operation	Month 1	2	3	4	5	6	7	8	9	10	11	12	13
1.	PRELIMINARIES.													
2.	ROADS + DRAINAGE.													
3.	SUBSTRUCTURE.													
4.	SUPERSTRUCTURE.													
5.	ROOF.													
6.	SERVICES.													
7.	INTERNAL FINISHES													
8.	EXTERNAL SITE WORKS													
9.	LANDSCAPE + CLEARANCE													

Fig. 9.4.1 The tender programme may have to be more detailed if the architect requests it

2. Concrete gang size.
3. Time of year that concrete is to be laid – winter working is less favourable.
4. Thickness and width of foundations – takes longer to lay 1 m^3 of concrete in thin foundations than in mass foundations.
5. Complexity of site.

Calculations

Foundation concrete is 180 m^3,
Can be laid at, say, 5 m^3/hour (this can be agreed with the Estimator or from Work Study records),
then 180 m^3 ÷ 5 m^3/hour = 36 hours,
say 8 hour working day,

then $\frac{36 \text{ hours}}{8 \text{ hours}} = 4\frac{1}{2}$ days

assuming 5-day working week,

then $\frac{4\frac{1}{2} \text{ days}}{5 \text{ days}} = \frac{9}{10}$ weeks, say 1 week.

Note: If more than one ready-mix truck has access to the foundation trenches and a further concrete gang was employed, this period could be reduced.

With the excavations, brickwork to dpc and concrete floor assessed to take: 10 weeks
Calculated duration for foundation concrete = 1 week

Total 11 weeks

The total duration for substructure work is 11 weeks and is lined in as shown in Fig. 9.4.1.

A similar procedure is followed for the other operations, the results of which are shown on the Tender Programme in a similar manner to the substructure work.

Chapter 10

Contractor's pre-contract work

10.1 Planning work

Having submitted a successful Tender a contracting firm needs to organise itself quickly to be in a position to commence work on-site on the date expected by the architect. The preliminary decisions taken at the Tender stage would be adjusted if necessary and finalised to enable the firm to at least take possession of the site. Taking possession without previous planning could be both costly and wasteful in time caused by delays in obtaining the necessary approvals, licences, permits, labour, materials, equipment and information on time. Prompt attention to such details are essential for the smooth commencement, running and completion of a project.

The stages of work undertaken by management in setting up the site with the correct requirements are discussed at a formal pre-contract meeting of the managers and senior personnel who will be closely associated with the project. Responsibilities will be delegated and the lines of communication with the architect and his team will be outlined to those present at the meeting; the outside lines of communication having been established at a meeting held with the architect and his representatives shortly after the contract has been awarded.

Those delegated with responsibilities can proceed with the preparation of the following major areas of pre-contract organisation:

1. Method Statement – see Section 10.2.
2. Contract Programme – see Section 10.3.
3. Site Layout Plan – see Section 10.4.
4. Schedules and other requirements – see Fig. 10.1.1.

Schedules and other requirements

Schedules

On completion of the Method Statement, Contract Programme and Site Layout Plan, numerous Schedules can be prepared so that action can be taken to proceed with the preparation for the commencement of work on-site. The schedules are essential to prevent anything from being overlooked which would prove undesirable later. The most common schedules are as follows:

1. Plant schedules (a) Mechanical: The decision to use the firm's own plant or to hire would have been taken at the Method Statement stage and by checking the Contract Programme a schedule of requirements can be made to enable an order to be placed in the right quarters, i.e. the firm's own plant department or with a plant hire firm.

(b) Non-mechanical: Scaffolding, hutting and other accommodation, furniture, cooking appliances, barrows, ladders, tarpaulins and many other items would be the subject of a further schedule.

The purchasing officer or contracts manager normally finds that the responsibility for requisitioning such plant items is his own.

2. Site Supervisors and Administration Staff Schedule: This schedule would be finalised and simultaneously a proper Site Organisation Structure Chart could be prepared to give the optimum supervision on-site and to enlighten everyone concerned with the responsibility of the site supervision and administration team. The chart would show the channels of communications clearly (see example in Fig. 3.4.1).

3. Labour Requirements Schedule: This is best prepared from the finalised Method Statement, Calculations Sheet and Contract Programme. The period for which the labour is required should be noted to assist the personnel officer or contracts manager in deciding whether the personnel is to be obtained by transference from other sites, or from sources outside the firm, i.e. job centres and from general advertising.

4. Schedule of Subcontractors' Work (a) Nominated subcontractors: Their names and trades are extracted from the Specifications and Bills of Quantities, and, either by arranging meetings with them, or waiting until being introduced by the architect at his pre-contract meeting, discussions must take place to agree on how they will fit into the contractor's programme.

(b) Contractor's own subcontractors: The firms who quoted the lowest subcontracting Tenders would now be notified, confirming the award of the

Information required schedule

Ref.	Detail of information required	Location	Date required	Comment	Date received	Comment

Plant schedule

Ref.	Description of plant	Quantity	Date to order	Date on site	Date off site	Order information		
						Supplier	Order no.	Order date

Labour schedule (own)

Ref.	Labour required	Number	Date on site	Date off site	Period	Supply information	
						Supplied from	Date

Subcontract schedule

Ref.	Description of material	Enquiry date	Order date	Date on site	Period	Subcontract details		
						Subcontractor	Order No.	Order date

Material schedule

Ref.	Description of material	Quantity	Enquiry date	Order date	Date on site	Supply details		
						Supplier	Order No.	Order date

Fig. 10.1.1 Examples of main contractor's pre-contract schedules of requirements

contract. Arrangements would next be made to bind them to the subcontract through the BEC Standard Form of Subcontract. Agreement is necessary on the dates the subcontractors are expected to do their parts of the work.

5. *Materials Schedules* (a) Nominated suppliers: Preliminary orders for critical items, i.e. special facing bricks or materials in short supply and long delivery dates, having been made by an efficient architect, need to be confirmed, and agreement for the dates of delivery have to be made with the suppliers.

(b) Contractor's own suppliers: From the schedules of the quotations received at the tender stage, the lowest with the most favourable terms would be chosen by the contractor's purchasing officer. Further schedules would also be necessary of the less critical materials so that orders can be placed.

6. *Schedule of Detail Drawings Requirements:* The most urgent Detail Drawings would be noted, and the dates by which they are required forwarded to the architect without delay.

Other arrangements

Many arrangements have to be made with outside parties before work commences on-site particularly in order to comply with the law. These arrangements are well known in construction work and a well-organised firm has Check-Lists to work from to prevent details from being overlooked. Some of the outside arrangements are outlined as follows:

1. Insurances – the type which forms part of the contract agreement as stated in Clauses 20, 21 and 22 of the JCT Standard Form of Building Contract Private Edition, with Quantities. Under the Employer's Liability (Compulsory Insurance) Act 1969, an insurance Certificate must be displayed on-site to show the employees they are protected in the event of an industrial injury being sustained.
2. Licences – obtained from the local highway authority for the erection of hoardings and gantries where they encroach upon a highway. Other licences are required from the local authority if it is intended that the storage of petroleum spirit or explosives is necessary on-site.
3. Permits. Skips: application made to the local highway authority under the Highways Act 1980.

   Overtime: where overtime working is anticipated, application for a permit should be made under the Working Rule Agreement, to the Local Joint Overtime Committee of the National Joint Council for the Building Industry. The permit should be displayed on-site.
4. Rates. These will be charged by the local authority where it is intended that huts will be used continuously for more than twelve months on-site. The charge by some authorities will not become due until after the twelve-month period and will be back-dated, but with other authorities rate charges commence immediately the job starts.

5. Public undertakings. Existing services: if necessary, information should be obtained of the positions of old, existing services under the ground from British Telecom, water, gas and electricity undertakings (also local authority engineer, regarding drains and sewers), to help prevent damage or injury during excavation, piling and other similar work.

   New Services: It is necessary to arrange for establishing temporary facilities on-site for building operations and obtain approvals for connecting to existing service mains (unless the undertakings are acting in a subcontracting capacity).
6. Commencement Notices. Local Authority: notice as laid down in the Public Health Acts and Building Regulations is required 24 hours before work commences on-site.

   Factory Inspectorate: where operations are to last for more than six weeks, notification is necessary. See Section 7.4 regarding notification to HM Factory Inspectorate.
7. Meals on-site – an application for permission to provide meals on-site should be made to the Department of Health and Social Security.
8. Closure or Part Closure of Highways – a permit is necessary from the local Highways Authority. Where the provisions of cross-overs are anticipated, a permit is also required.
9. Adjoining property – where there is a likelihood of infringement of an adjoining property owner's air space when using a tower crane, approval should be sought from the owner to prevent the possibility of court action later. It would also be prudent to agree with the adjoining property owner any damages, however small, which exist before work commences on-site – tell-tales and photographs would assist here as evidence.
10. Police – It is usual to notify them of impending site work, it not only helps with security later, but is also courteous.
11. Trade Unions – should be forewarned of new site works (they can help in recruitment) to assist in their administration work.

There are many arrangements to be made by the contracting firm from within its organisation, a few of which are listed here:

1. Notifications and Agenda preparation for a pre-Contract Meeting.
2. Bonus and overtime arrangements.
3. Transport facilities prepared.
4. Health and Welfare facilities.
5. Safety provisions.
6. Security arrangements (possibly assisted by Consec which is a subsidiary of the Building Employers Confederation (BEC)).
7. Prepare site notice board, if new one required.
8. Arrange samples to be provided to the architect from suppliers where necessary.
9. Check list of relevant site documents, stationery, forms, registers, etc. and submission of a requisition.

10. Arrange transfer of site staff and operatives, including provisions for salaries and wages.
11. Cost Budget and arrangements for financing the work.

One hopes that on commencement of work on-site few, if any, extra provisions have to be made available in addition to those planned at the pre-contract period.

10.2 Method statement

This document is best prepared at the earliest opportunity on the award of a contract. It is used as a guide to everyone within the contracting firm regarding the method and sequence of construction.

During the preparation of the Method Statement the construction work should be broken down into operations, and the labour and plant requirements decided upon by those responsible for the documents' preparation, i.e. planning/contracts manager, programming officer, estimator and plant manager. The Pre-Tender decisions would be used as a guide, perhaps with some adjustments in the light of new developments.

The Method Statement is a basis from which the programming officer works while drawing up the Contract Programme. It is also used later on-site by the site manager and should convey easily how it is intended that the work should proceed, particularly where unusual construction techniques are envisaged.

Many firms forgo the preparation of such a document and rely entirely on the site managers to carry out the work to the best of their ability and experience. In repetitive construction work, particularly housing, little use can be found for a Method Statement, but it is expected that at least a pre-contract meeting is held by all those concerned with a project to agree on even the simplest method and sequence for doing the work.

During the Method Statement preparation (see Fig. 10.2.1), due regard is paid to the contract documents, particularly the Specifications, Drawings and Bills of Quantities, and if the work is to be handed over in stages, it would be included in the Method Statement.

All decisions taken at this stage of the pre-contract work should be fairly conclusive and once the Method Statement is finalised, agreement should be sought from those who prepared it before deviations are made from the decisions so incorporated.

At the pre-tender stage, decisions would have been made on the work which is to be executed by the contractor's own subcontractors, and, along with the Nominated Subcontractors, they would be incorporated in the Method Statement where required.

10.3 Contract programme - Bar Charts/Critical Path Charts

A normal prerequisition to the preparation of a Contract Programme is that the duration of each of the major operations outlined in the Method Statement is calculated. The extent of the work involved for each operation can be established by studying the Contract Documents in depth. Quantities can be extracted from the Bills or by scaling off from the Drawings, and as the calculated data will be incorporated eventually in a Bar Chart, Critical Path Chart, Line of Balance Chart or Precedence Diagram, the final number of major operations should be restricted to between twenty and fifty, depending on the size and complexity of the contract work.

If the Two Storey Office Block was once more considered, as in the previous chapter, particularly under Tender Programme (Fig. 9.4.1), it is assumed that the work can be enlarged further into the operations as shown in a typical Calculation Sheet (see Fig. 10.3.1). This Calculation Sheet is used for establishing operation times/durations with adequate precision.

However, for some contracts, an experienced programmer can estimate how long operations are going to take, and by adding them can arrive at a contract duration which may not be too far removed from that obtained by calculations. If the programmer were to prepare the contract duration by the two techniques, at least this form of collaboration would serve as a check on the duration in addition to the decisions made on the Tender Programme.

If the contract duration is at variance to the Tender Programme, adjustments can be made to selected operations by the following:

1. Use of larger gang sizes.
2. Use of additional gangs.
3. Introduction of a sound bonus scheme.
4. Allowing overtime.
5. Use of bigger/or more efficient mechanical plant.
6. Introduction of better methods and techniques of construction.
7. Employment of more supervisors to control and monitor the work.

Calculation Sheet preparation (see Fig. 10.3.1)

1. Operations: These should have been decided at the Method Statement stage and should be listed on the Calculation Sheet.
2. Quantities: Where Bills of Quantities form part of the contract documents, quantities can be extracted. In other cases, scaled from the drawings.
3. Average rate/hour: should be obtained from Work Study records or from knowledge gained through past experience by the programmer. The rate is the volume of work which can be done by the principal operative (joiner, bricklayer, etc.), or in, say, bulk excavation – the excavating plant.
4. Total hours: Divide the average rate/hour into the quantities.
5. Plant/gang size: by experience a choice can be made, but where the Method Statement has been prepared, the type of plant and gang sizes can be extracted.
6. Duration in hours: obtained by dividing the principal operations shown in the plant/gang size column into the total hours.
7. Duration in days: by converting the duration in hours into days for ease of plotting on the Contract Programme.

Method statement

Contract: Two Storey Office Block. Sheet no: 1 Date: 10th Aug. 19

Client: S. Helen Ltd. Prepared by: S. Mavrs / G. Bacs.

Op. no.	Operation	Method and sequence	Plant	Labour	Notes
1.0	Prelims etc.				
	1.1 Access.	See site layout plan for access position. All rubbish to be removed from site and should not be burnt.	J.C.B. 7B	3 lab.	Usual selection of bucket sizes required.
	1.2 Huts.	See site layout plan for positions and storage of materials. Huts to be erected on brick piers - clean up and decorate before use.	-	3 lab.	-
	1.3 Temp. Services.	Standpipe as on site layout plan. Telephone into Gen. Foremans office only. Mess huts and Gen. foremans Hut to have gas and electric points. Connect W.C. to mains at earliest opportunity.	-	Sub. Con.	Protect all pipes and wires.
	1.4 Site Clearance.	Commencing at main road road end and steadily progressing to far end of works, loading onto lorry and cart to tip at Summerhill, London Rd.	J.C.B. 7B	3 lab.	Use J.C.B. 7B for grubbing up roots.
	1.5 Setting out	Drains and roads first then main building	-	Gen. F/man 3 lab.	New supply of pegs & boards.
2.0	Drains & Roads				
	2.1 Exc. Drains.	Work out towards main road and excess excavated material deposited in spoil heap awaiting cartage to tip.	J.C.B. 7B	3 lab.	Notify building inspector at each stage of work.
	2.2 Drain Connection to Sewer	At end of drain works.	-	Sub. Contractor	
	2.3 Excavate to Reduced Road Level.	Start from main road and continue without interuption until completed Excess material removed to tip.	J.C.B. 112 2 lorries	3 lab.	
	2.4 Road Hardcore	Start as above and consolidate by vibrator roller.	Vibrator roller	3 lab.	
	2.5 Kerbs	Complete all kerbing ~~before~~ concreting road surface	170L Conc. Mixer	3 lab.	
	2.6 Conc. Road.	Start at top end of site using ready-mix concrete and work out to main road.	Vibrator tamper	6 lab.	Protect surfaces at end of each working day.
	Student note:	The remaining operations are listed and descriptions continued for them similar to the aforementioned.			

Fig. 10.2.1

Student Note: The remaining operations are listed and descriptions continued for them similar to the aforementioned

Calculation sheet

Contract: Two Storey Office Block — Sheet No: 1 — Date: 29th Jan 19

Client: B Helen Ltd — Prepared by: S Morrs

Op No	Operation	Quantities	Average rate/hour	Total hours	Plant/gang size	Duration in hours	Duration in days
1.0	Preliminaries						
	Access	–	–	–	JCB 7B 3 lab	–	2
	Huts	3 Huts	–	–	3 lab	–	8
	Temporary Services	–	–	–	Sub Con	–	3
	Site Clearance	–	–	–	JCB 7B 3 lab	–	2
	Set-out Works	–	–	–	F/Man 3 lab	–	5
2.0	Drains and Roads						
	External main Drains	400 m	1.2 m/hr	334 hrs	JCB 7B 3 lab	112	14
	Connect Drains to Sewers	–	–	–	Sub Con	–	3
	Excavate to Road Formation Level	1800 m²	30 m²/hr	60	JCB 112 3 lab	60	8
	Hardcore to Roads	1800 m²	8 m²/hr	225	Vib. Roll. 3 lab	75	10
	Kerbs	650 m	4 m/hr	163	170 L. Conc Mix. 3 lab	55	7
	Concrete Road	1800 m²	5 m²/hr	360	6 lab	60	8
3.0	Sub-Structure						
	Excavate Foundation Trench	320 m³	10 m³/hr	32	JCB 7B 1 lab	32	4
	Concrete Foundations	180 m³	1.5 m³/hr	120	3 lab	40	5
	Brickwork to D.P.C	300 m²	1 m²/hr	300	2 b/l 2 lab + 100L mixer	100	13
	Internal Drains	150 m	1 m/hr	150	3 lab	50	7
	Hardcore to Floor	700 m²	6 m²/hr	117	3 lab	39	5
	Concrete Ground Floor	700 m²	4 m²/hr	175	6 lab	28	4
4.0	Superstructure						
	Brickwork and Blockwork	3000 m²	1 m²	3000	9 b/l 6 lab 100L Mixer	334	42
	Formwork to Conc. 1st Floor + Stairs	800 m²	3 m²/hr	267	3 C+J	89	12
	Reinforcement to Conc " + Stairs	–	–	–	Sub. Con	–	10
	Concrete to 1st Floor + Stairs	700 m²	3 m²	234	6 lab	39	5
	External Door and Window Frames	–	–	–	Sub. Con	–	8
5.0	Roof						
	Structure (Timber)	–	–	–	3 C+J	–	12
	Tiling	850 m²	1.5 m²/hr	567	Sub. Con	–	20
	Rainwater Goods	–	–	–	Sub. Con.	16	2
6.0	Services						
	Plumbing and Heating	–	–	–	Sub. Con.	–	55
	Gas	–	–	–	Sub. Con	–	10
	Electric	–	–	–	Sub. Con	–	40
	G.P.O	–	–	–	Sub. Con.	–	15
7.0	Finishes						
	Carpentry and Joinery	–	–	–	3 C+J	–	60
	Plastering and Screeds	–	–	–	Sub. Con	–	44
	Floor Finishes	–	–	–	Sub. Con	–	20
	Decorating and Glazing	–	–	–	Sub. Con.	–	40
8.0	External Works						
	Footpaths	250 m²	2 m²/hr	125	170 L. Conc Mixer. 3 lab	42	6
	Landscaping	–	–	–	3 lab	–	15
	Clearance	–	–	–	3 lab	–	15

Student note: The quantities are taken from the drawings and bills of quantities. Average rate/hour obtained from past records or experience by the programmer. Only critical major items require calculations for durations and associated work must be taken into consideration i.e. level and ram bottom of excavation included with excavation of trench

Fig. 10.3.1

Student Note: The quantities are taken from the drawings and bills of quantities. Average rate/hour obtained from past records or experience by the programmer. Only critical major items require calculations for durations and associated work must be taken into consideration, i.e. level and ram bottom of excavation included with excavation of trench

Contract Programme

A bar chart or Gantt chart is the commonest and most popular method of illustrating the sequence of operations of a contract and the expected durations. Its popularity stems from the fact that the chart is simple to understand and use.

This control document is suitable to instruct the site manager, in particular, the order in which operations are to be executed. By referring to the document, a plan of action can be formulated in order to carry into operation the planner's/programmer's overall plan, and in the time allowed.

Many site managers complain that the programme soon becomes unusable because of delays caused by inclement weather, late material deliveries, subcontractors and instructions not being given on time by the architect. When problems such as these develop, the programme should be up-dated to show a revised plan of action.

Site managers argue that many durations are unreasonable and should be longer. This is true in many cases, but at least corrective action can be taken on subsequent operations by allowing overtime working and by introducing additional labour and plant on to the site.

Without the programme, the site manager would encounter difficulty in assessing whether work is behind or ahead of schedule; therefore the old saying of 'a bad plan is better than no plan at all' is very true.

Bar Chart preparation

The essential details, i.e. operation numbers, operations, plant/gang sizes and durations in days, are extracted from the Calculation Sheets on to a blank prepared standard copy of a Contract Programme. The programming officer next begins to draw in (in pencil at first) the duration for each operation, so that an operation follows in sequence and as early as possible after the previous operation. The programmer must then consider the following if he/she is to achieve a satisfactory programme bar chart:

1. Certain small operations are not included on the chart but adjustments must be made to the operations most affected by them, i.e. the level and ram bottom of excavations' duration should be added to the excavation for foundation trenches. Also, the duration for laying DPCs on to brickwork would be included with brickwork in the Superstructure Section.
2. Some operations need not be delayed until a previous critical one is completed; for example, the concrete to ground floor can commence before all the hardcore is completed. This assumes that there will be two labour gangs available, one for the hardcore and the other for concreting the floor.
3. Continuity of work for labour and plant should be the aim but is not always possible. An attempt must be made while drawing in the horizontal duration bars on the chart that, say, after the JCB 78 has completed the external drainage, it immediately progresses on to digging foundation

trenches (see Fig. 10.3.2), this prevents idle time and unnecessary expenditure on the contract.

4. If after drawing a duration bar for an operation it appears excessive – especially where it is obvious that other operations will be seriously delayed – by increasing the gang size the bar could be reduced to a more realistic length (duration), thereby enabling the contract to be completed earlier.
5. Correct gang balance is important for continuity of work. The numbers of operatives on-site should not fluctuate too dramatically, and one should allow for the numbers to increase gradually early in the contract, then to a peak mid-way through and gradually reducing at the tail-end of the contract. This primarily depends on whether most of the work is executed by the contractor's operatives or by subcontract labour.

If there is a heavy 'liquidated and ascertained damages' clause in the contract, a contractor would be prudent to fix himself/herself a Target Date for completion of the works, which may be from one to six weeks before the client's date of possession so that in the case of the Target Date being optimistic at least he/she would have endeavoured to complete the contract and it is hoped then that the date of possession deadline is at least reached.

For a typical Contract Programme which shows most details for conducting and controlling the work on-site, see Fig. 10.3.2. Where it is expected that the site manager has the responsibility for chasing special orders for materials, giving notifications to subcontractors and calling for information from the architect and design team, special symbols are included on the bar chart on the dates action should be taken.

Critical Path Charts or Critical Path Method (CPM)

As an improvement and more precise system of programming than the Bar Chart, the CPM or Network Charts were introduced. This form of chart shows which activities or operations are more important than others, i.e. those on the critical path. Those activities which are most important are generally the ones that need to be completed on or before the end of the durations allocated to them. If an activity takes longer than the time allowed, it would have the effect of extending the contract period with all the normal ramifications, i.e. disorganisation, cost increases and displeasure of the architect and client.

Network preparations

In the same way as Bar Charts are initially dealt with, the preparation of networks begins by:

i. Listing the activities/operations of the proposed project in the approximate order of execution.
ii. Calculating the durations of each activity.

Now continue the preparation of the network (see Fig. 10.3.3) as follows:

iii. Sketch out a Network Chart in a logical way after first asking the question, 'What operations can be done first or simultaneously with others?' – avoiding the question 'What should be done first?' (as in Bar Chart preparation). Do not even consider labour plant or other resources at this stage.

iv. Add to the chart the durations of each activity.

v. Calculate the earliest and latest start and finish times and critical path(s) as follows:

Earliest start times: Activity 1–2 = beginning of first day, hence, day 0.
Activity 2–3 = day 0 + 6 days' duration of 1–2 = 6 days.
Activity 2–5 = 6 days.
Activity 2–4 = 6 days.
Activity 3–5 = 6 days + 5 days' duration = 11 days.
Activity 3–6 = 11 days.
Activity 4–7 = 6 days + 15 days = 21 days.

For the earliest start time of activity 5–8 there are two routes into node 5, therefore, the 'longest route' 1–2 to 2–5 = 20 days is taken in preference to route 1–2 to 2–3 to 3–5, which only adds up to 15 days. Naturally, activity 5–8 cannot start until activity 2–5 is completed at 20 days. This method applies where more than one arrow/activity enters a node, as at node 6, node 7 and node 8.

Latest start times: These are calculated starting at the earliest completion day, i.e. day 34. Work backwards choosing the 'shortest route' back along the network. For further details refer to Fig. 10.3.3.

10.4 Site layout

The most convenient method of instructing a site manager as to the best positions for hutting, temporary service points, equipment and materials is by a site layout plan drawn to a suitable scale (see Fig. 10.4.1). One attempts to cause as few obstructions to the work as possible, but where careless sighting of huts and materials have taken place, the smooth running of construction work could be seriously interrupted by unnecessary double handling of materials or repositioning of huts later to enable certain work to progress. Well thought-out site layout plans help a contractor to start work on-site on the right foot leading, it is hoped, to unhindered progress throughout.

By using scale drawings (normally traced from the architect's block or site plans), the planner/programmer, assisted by the contracts manager, prepares a site layout plan by drawing in the proposed administration and other huts including positions for the storage of special or bulk materials, etc. In preparing positions, the planner should take cognisance of the setting out grid (if known); proposed building; new services and roads; and areas the client has requested should not be encroached on.

Contract: Two Storey Office Block.
Job No: C/22/77
Date: 11th. Aug. 19

BUILDING CONTRACTORS LTD

Contrac

Month: FEB 19 | MAR. | APR. | MAY.

Week commencing	7	14	21	28	7	14	21	28	4	11	18	26	2	9
Wk. No.	1	2	3	4	5	6	7	8	9	10	11	12	13	14

Op. No.	Operation	Gang plant size	Duration (in days)
1.0	PRELIMINARIES		
1.1	Access	JCB 2 lab	2
1.2	Huts	2 lab	8
1.3	Temp. Services	Sub Con	3
1.4	Site Clearance	JCB 3 lab	2
1.5	Setting Out	F/Man 3 lab	5
2.0	DRAINS & ROADS		
2.1	Ext. Main Drains	JCB 3 lab	14
2.2	Connect Drains to Sewer	S.C.	3
2.3	Exc. to Road Form. Level	JCB 3 lab	8
2.4	Hardcore to Roads	VIB ROL 3 lab	10
2.5	Kerbs	Mixer 3 lab	7
2.6	Conc. Road Surface	6 lab	8
3.0	SUBSTRUCTURE		
3.1	Exc. Found. Trench	JCB 1 lab	4
3.2	Conc. Found.	3 lab	5
3.3	BWK. to D.P.C.	Mixer 3 bk. 2 L	13
3.4	Int. Drains	3 lab	7
3.5	Hardcore to Floors	3 lab	5
3.6	Conc. Ground Floor	6 lab	4
4.0	SUPERSTRUCTURE		
4.1	BWK & Blockwork	Mixer 9 bk 6 L	42
4.2	Formwork to 1st Floor	3 C&J	12
4.3	Reinf. to Conc. " "	SC	10
4.4	Conc to 1st Floor	6 lab	5
4.5	Ext. Door & Wind. Frames	SC	8
5.0	ROOF		
5.1	Structure (timber)	3 C&J	12
5.2	Tiling	SC	20
5.3	Rainwater goods (etc)	SC	2
6.0	SERVICES		
6.1	Plumbing & Heating	SC	55
6.2	Gas	SC	10
6.3	Electricity	S.C.	40
6.4	G.P.O.	S.C	15
7.0	FINISHES		
7.1	Carp. & Joinery	3 C&J	60
7.2	Plaster & Screeds	SC	44
7.3	Floor finishes	SC	20
7.4	Paint & Glazing	SC	40
8.0	EXTERNAL WORKS		
8.1	Footpaths	Mixer 2 lab.	6
8.2	Landscaping	2 lab	15
8.3	Clearance	3 lab	15

2.1 | 3.1 | 4.1 | BANK HOLIDAY

		Wk 1	Wk 2	Wk 3	Wk 4	Wk 5	Wk 6	Wk 7	Wk 8	Wk 9	Wk 10	Wk 11	Wk 12	Wk 13
LABOUR:	Gen. Lab.	3	6	6	6	6	6	6	6	6	6	6		
	Bricklayer						3/2	3/2	3/2	3/2	9/6	9/6	9/6	9/6
	C & J													
PLANT:	2 Hoists													
	J.C.B. 7B													
	J.C.B. 112													
	Scaffold													
	170L Conc. Mix													
	100L Mortar Mix													

Client: B. Helen Ltd. Greystone Rd. Cambs.
Architect: D. Joanne A.R.I.B.A.
Q.S.: M & M. Hay A.R.I.C.S.

Symbols:
- ● Delivery.
- ◆ Information.
- → Notification.

Fig. 10.3.2

ERROW, GUILDFORD 2657.

rogramme

Prepared by: S. Morrs.

			JUN.				JUL.				AUG.					SEPT.				OCT.					NOV.				DEC.
16	23	30	6	13	20	27	4	11	18	25	1	8	15	22	29	5	12	19	26	3	10	17	24	31	7	14	21	28	5
15	16	17	18	19	20	21	22	23	24	25	26	27	28	29	30	31	32	33	34	35	36	37	38	39	40	41	42	43	44

SPRING BANK HOLIDAY

LATE SUMMER HOLIDAY

CONTRACT COMPLETION

5.1

1st FIX

2ND FIX

6.1

1st FIX

2ND FIX

7.1

8.1

6 6 ... 3 3 6 6 6 ----

9/6 ------ 9/6 9/6 9/6 9/6

3 3 3 3 3 3 3 3 3 3 3 3 ------ 3 3 3 3 3 3

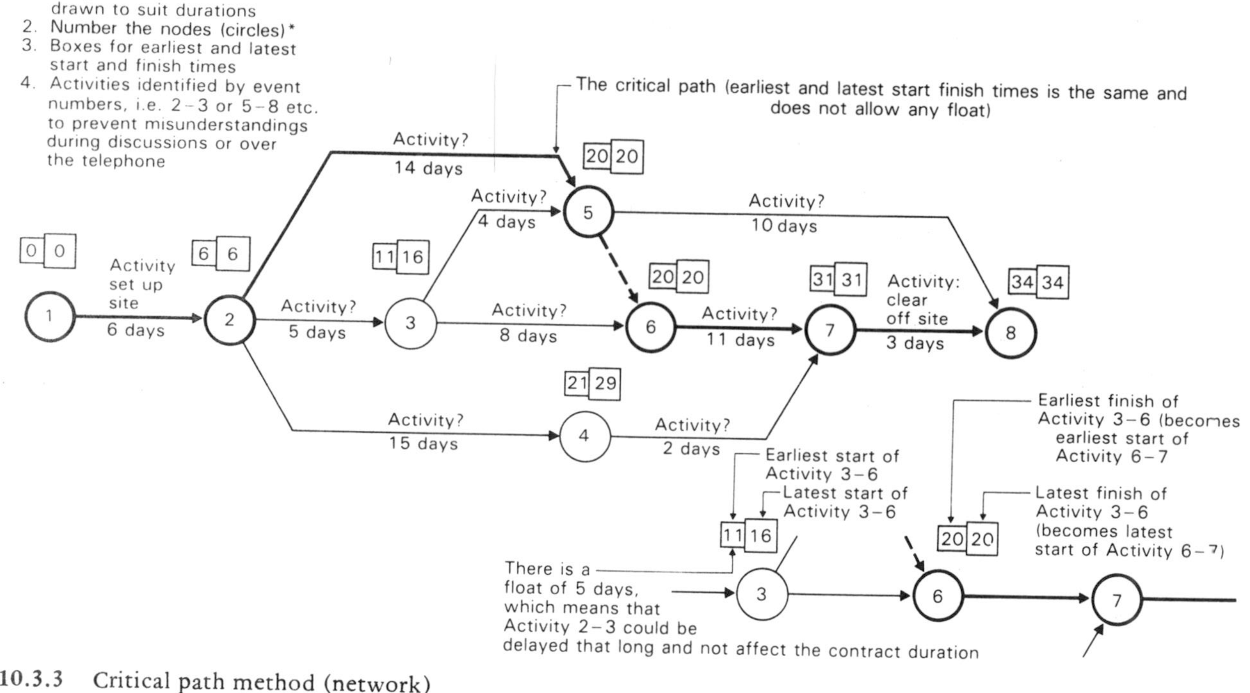

Fig. 10.3.3 Critical path method (network)

Note:

1. Length of arrows not drawn to suit durations.
2. Number the nodes.
3. Boxes for earliest and latest start and finish times.
4. Activities identified by event numbers, i.e. 2–3 or 5–8, etc. to prevent misunderstandings during discussions or over the telephone.

While bearing in mind the complexity and size of the works, both horizontally and vertically, the following are suggested topics to be carefully considered, some of which are necessary under the Working Rule Agreement and Construction Regulations:

1. Scaffold – draw this in relative to the building which could then determine the hoist or tower crane positions.
2. Site Manager's and other administration huts – the ideal position for good supervision would be the one which gives an all-round view of the site, messrooms, delivery points and vehicle parks. It should have separate toilet facilities.
3. Clerk of Works' hut – provided close to site manager's hut.
4. Compound and workshop – to store valuable items, and should be made secure with high wire mesh and adequate gates, chain and lock. Materials to be stored away from the boundaries both within and without.

   Racks for scaffolding and timber to be preferably provided, with lock-ups for fuel and valuable moveable items. It is ideal to have a storeman responsible for all material issues and returns.
5. Mess huts – site conditions are enhanced if provisions are made for hot meals, sometimes by subcontract canteen staff. On small sites, however, at least facilities should exist for the preparation of hot drinks.
6. Adjoining the mess hut should be a drying room for workmen's clothing.
7. Parking area for vehicles – workmen and visitors to site are less frustrated by making this available. However, so that the temptation to pilfer is lessened, this area should not be adjacent to the mess hut or the work area.
8. Storage areas – this should be marked out clearly for the varying materials (see Fig. 10.4.1).
9. Mixer positions – a hardstand for mixers and aggregates helps prevent waste, with a further provision for storing cement, etc.
10. Standpipes – close to mixer positions or where required most.
11. Electricity transformer point – centrally positioned as agreed with Electricity Board.
12. Site name board – in the most prominent position from the highway.
13. Lighting – for security reasons at night a well-lit site discourages thieves. The lighting should be positioned near to stores and other less secure areas.
14. Weighbridge – positioned close to the checker's office (if project is big enough to warrant one).
15. Vehicle brush and wash area – helps to prevent dirty highway in the immediate vicinity of site.
16. Skip position – should be close to the works to help assist in site cleanliness (should be emptied at regular intervals).
17. Hoardings, gantries and pedestrian passageways – carefully detail on drawing when required.
18. Tower crane – this should be drawn close to the proposed building to show its travel distance and boom swing. When drawn to scale, its most efficient working area is determined.

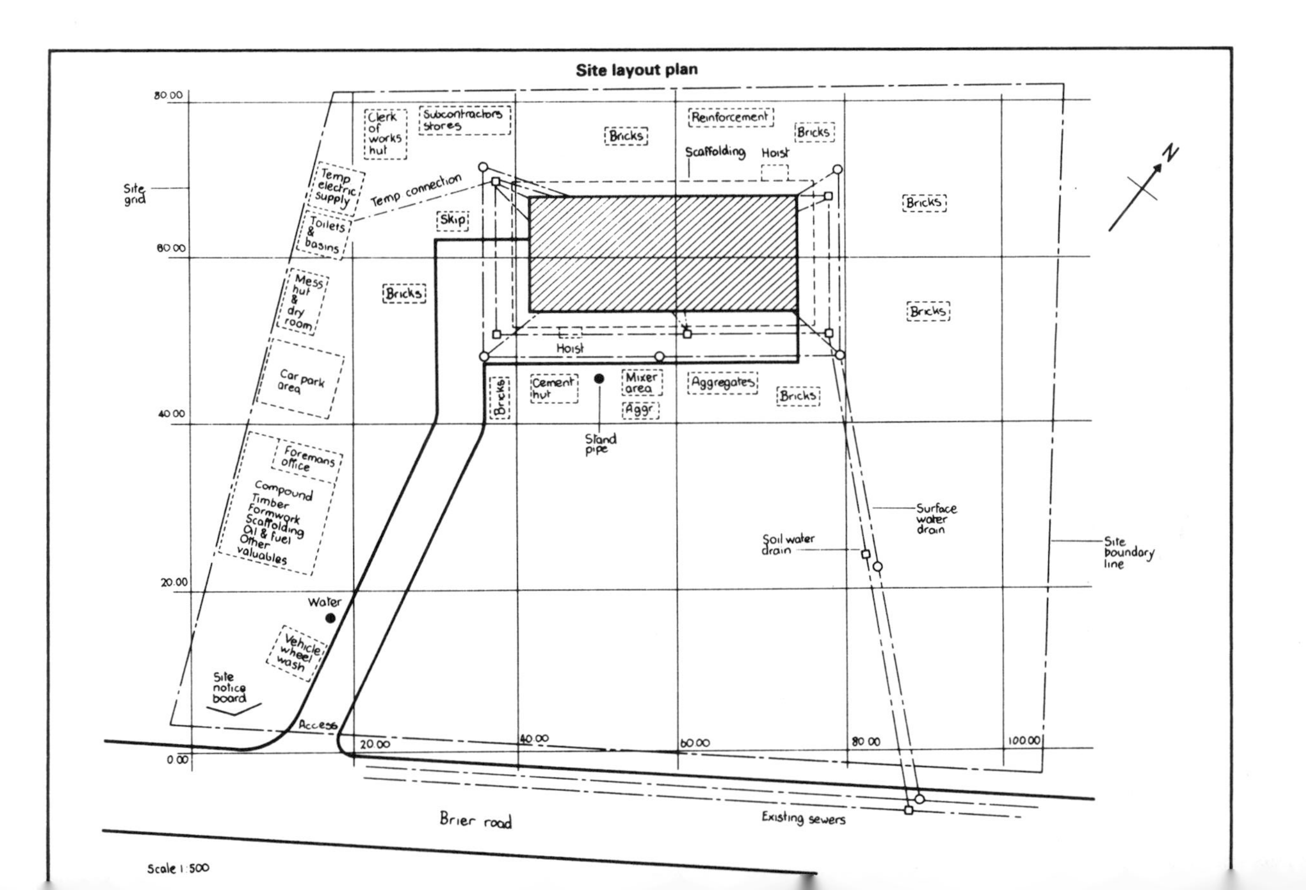
Site layout plan
80.00
60.00
40.00
20.00
0.00
20.00
40.00
60.00
80.00
100.00
Site grid
N
Clerk of works hut
Subcontractors stores
Bricks
Reinforcement
Bricks
Scaffolding
Hoist
Temp electric supply
Temp connection
Toilets & basins
Skip
Bricks
Mess hut & dry room
Bricks
Bricks
Hoist
Car park area
Bricks
Cement hut
Mixer area
Aggr
Aggregates
Bricks
Stand pipe
Foremans office
Compound Timber Formwork Scaffolding Oil & fuel Other valuables
Surface water drain
Soil water drain
Site boundary line
Water
Vehicle wheel wash
Site notice board
Access
Brier road
Existing sewers
Scale 1:500

19. Passenger lifts – should be located close to operatives' easiest access.
20. Other considerations, such as: direction finger boards for merchants' and others' deliveries to site, and to prevent vehicles straying on to adjoining premises or roads which could lead to bad public relations; stop-go lights to prevent congestion on some sites; position of vegetable soil spoil heap; viewing platform for the inquisitive public; underground existing service positions to prevent digger damage.

In preparing for a well-laid out tidy site, a contractor helps to convey an efficient organised image to the general public which can only enhance his/her position.

Chapter 11

Contract work and other considerations

11.1 Setting up site and controlling works

It is now assumed that all relevant Notices, Permissions and Licences have been given or obtained prior to commencement of work on-site by those responsible for doing so at the contractor's head office. In taking possession of the site, the project/site manager has many tasks to perform, for example:

1. The setting out of the access position for the excavator driver (if necessary), using either the contract drawings or site layout plan which was prepared at the pre-contract period.
2. The preparation of a site grid of, say, 15 to 20 metres square which can be used to locate the building position on the land. Pegs are positioned round the boundary of the site and a theodolite is used to determine where a proposed manhole, corner of the building and point on a road should be placed (see site layout plan, Fig. 10.4.1, which should be to scale). At intersections of the site grid close to the boundaries or other convenient positions, pegs should be inserted. It is over these pegs that a theodolite is set up and orientated, and by measuring along the theodolite line of sight (the length of which is predetermined), relative positions of the building works can be established.
3. The recording of site levels from a Temporary Bench Mark (TBM) which

is secured by a series of flying levels, if necessary, from an Ordnance Bench Mark (OBM).

4. Establishing the boundaries of the site, the building line, and improvement line (to be checked by the local authority).
5. Setting out the positions for hutting, compounds, storage areas, standpipes, etc.
6. Fixing of danger signs for overhead lines and underground cables and service positions to minimise damage.
7. The actual setting out of the building, drainage and roads, using pegs, profiles, sight rails, batter rails, etc.

Preliminary requirements

The contract manager or some responsible person at head office would have by now requisitioned all the preliminary requirements for the site, namely:

1. Hutting and ancillary equipment (chairs, desks, etc.). Exact requirements are extracted from the Method Statement, Schedules, Contract Programme and the company's standard check-list.
2. Administration Documents. Stationery, Prescribed Forms and Registers, Regulations (see Section 8.4).
3. First Aid Box (or facilities). The contents depend on the optimum numbers of personnel expected on the site during the contract. As a reference and to conform to the minimum requirements, see the Construction (Health and Welfare) Regulations 1966 and the Health and Welfare (First Aid) Regulations 1981.
4. Materials. Preliminary list used for immediate requirements, and Schedules in other cases.
5. Plant. From Schedules, Method Statement or Contract Programme.
6. Small Plant and other Equipment. Standard check-list suitable.
7. Temporary Services. Subcontractors notified.

As preliminary requirements are delivered, the site manager should reconcile them with the delivery notes or check-lists which should be provided, and shortages should be reported.

Consideration should be given at this stage to maintaining good relations with the public and ensuring that complaints are minimised by keeping down noise and dust. Site traffic and operations cause numerous disturbances and every endeavour should be made to retain the existing environmental conditions.

Adequate preparation during the early days of a contract will pay dividends later. When the works have been accurately set out, arrangements should be made to ensure that there will be economy of labour by reducing double handling of materials. Careful stacking of materials close to where operatives will require them will lead to a more satisfactory situation, and by minimising double handling and transportation on-site will encourage operatives to increase output to the benefit of all.

The site manager should be prepared to delegate responsibilities to individuals at an early stage in the contract regarding the following:

1. Security – opening up and closing the site daily; the general foreman or ganger being suitable here.
2. Temporary Services Control – turning water, electricity, etc. on and off before and after weekends, holidays, or providing protection during low temperatures (Ganger).
3. Start and finish signals – sounding the whistle, bell and buzzer at starting and finishing times, and for tea and lunch breaks (timekeeper or general foreman could do this).
4. Cleanliness of huts – ganger could organise this on a once-weekly basis, but the canteen would be the canteen manager's responsibility.
5. Site tidiness – each trade foreman and gangers to be responsible for their own materials and equipment.
6. Holidays – it may be necessary for the site manager to prepare a rota to ensure that a reasonable level of supervisors and administrators are maintained during the peak holiday period, unless the site is to be closed down because too many operatives have requested leave, making it impossible to maintain a steady flow of work.
7. Protection of work – protection is necessary to safeguard against frost, rain, sun and the wind, and carelessness of operatives. Each trade foreman and ganger would be expected to protect their own operatives' work.
8. Safety inspections, excavations and documentations.

Control

Administration and site work during a contract can be divided into many important areas which require some degree of control, the ultimate responsibility for which lies with the site manager.

Communications control

1. Mail – The receipt and dispersal of the mail could be one of the administrative staff's responsibilities, i.e. timekeeper or clerk, and general headed mail should be opened and passed to the appropriate person (usually the site manager), and personal mail forwarded unopened. When correspondence has been dealt with, it should be filed correctly.
2. Drawings Registers – These Registers should be prepared at the beginning of the contract and need to be kept up to date, particularly when drawings are amended or superseded (see Fig. 11.1.1). Separate Registers are kept for drawings received from each of the following:
   (*a*) architect
   (*b*) engineer
   (*c*) services engineers.

BUILDING CONTRACTING LTD.

Drawings register

Contract: Two Storey Office Block.

Contract No: C/32/85

Drawings from: Architect.

Date commenced: 7th Feb. 1985

Drawing No.	Description	Scale	Number supplied	Date of issue	Amendments								File ref.
					A		B		C		D		
					No.	Date	No.	Date	No.	Date	No.	Date	
1.	Block Plan.	1:500	2	7th Feb. 85									Drawer No. 1.
2.	Soil + Surface Water Drainage.	1:250	2	12th Feb. 85									Drawer No. 2.
3.	Foundation Layout.	1:100	2	18th Feb. 85									Drawer No. 1.
4.													
5.	Roadworks (access).	1:250	2	18th Feb. 85	2	26th.2.85							Drawer No. 1.
6.	Elevations.	1:100	2	26th Feb. 85									Drawer No. 3

Note: This assumes that a separate register is being kept for drawings which are issued by the Architect, Engineer and Service Engineers respectively. Instead of using a file reference column an Index System could be introduced if hundreds of drawings are involved.

Fig. 11.1.1

3. Filing – Particular emphasis should be given to careful filing of correspondence, documents and drawings. A suggested system of filing is by alphabetical means with folders clearly marked with either the name or title of the supplier, manufacturer or individual. Subject filing can be used for filing documents alphabetically, such as: safety documents and records, and personnel records. Drawings can be filed in numerical order but under the headings: architect, engineer, and services engineers, plan chests or vertical storage being preferred.
4. Telephone – All important calls should be recorded, particularly if instructions are received and action needs to be taken. Private personal calls should only be made with prior permission from the site manager.

Operations control

1. Contract Programme – This should form the basis of controlling all operations on-site. The programme gives visual instructions of when operations begin and end. It warns the site manager about which operatives will shortly be required and in what numbers (see Section 10.3).
2. Method Statement – If this document has been thoroughly prepared, it will explain the method and sequence of construction, and will outline the plant, equipment and labour required at each stage of the work (see Section 10.2).

Quality control

By referring to the contract documents, particularly the Specifications, the quality of the work expected by the architect and client can be determined. Sometimes, in order to keep the price of a contract to a minimum, the specifications are relaxed and the contractor then works to the minimum requirements.

Trade foremen and gangers are responsible in the first instance to ensure that required standards are maintained by their subordinates, and it is then up to the site manager to check periodically that standards are reasonable.

When materials and components are received on-site, where possible, and before unloading takes place, the Checker or someone in authority should check on the quantity and the quality. Damaged and unsatisfactory goods should be rejected and sent back to the suppliers.

If the site manager and subordinates overlook quality control, an observant Clerk of Works (representative of the client) will reject any sub-standard work or materials, thereby causing the builder to incur further expenses in correcting faulty workmanship, etc.

Materials control

Invariably, the purchasing officer at head office ensures that materials are ordered on time and that stock levels are maintained on-site. However, if the site manager is given this responsibility, quantities are extracted from the drawings and either requisitions are prepared at suitable intervals, or orders are called forward by him directly from the suppliers, a general order having first been made at the pre-contract stage.

BUILDING CONTRACTORS LTD., MERROW, GUILDFORD

No: 3742

Materials received

Contract:

Contract No:

Week ending:

Date received	Order number	Delivery note number	Supplier	Materials		Invoice number	Invoice amount		Rate	Code	Cost	
				Description	Quantity							

Site manager or checker would fill in a line of the materials received form for each quantity received on site up to and including the invoice number column. The remainder is completed at head office by the purchasing department or accounts department.

Fig. 11.1.2

There are, of course, other considerations when controlling materials on-site, for example:

1. Chasing orders – if the site manager is given the responsibility for chasing orders, then it would be prudent for him to phone the suppliers a few days prior to the previously agreed delivery dates to confirm that deliveries will take place. If problems have developed at the suppliers' end, then alternative arrangements can be made in the short term.
2. Daily or weekly material returns – to assist head office in reconciling materials delivered to site with those ordered, daily or weekly materials received forms are filled in by the site manager or site staff from the delivery notes. Continuous checks can then be made of the material costs throughout a project period (see Fig. 11.1.2).
3. Storekeeping – all materials issued from the stores should be signed for. Levels of stock are controlled by the use of materials-issued and materials-returned books and stock cards. When deliveries are made by suppliers, careful checking of quantities by storeman/checker will help to reduce the possibility of short deliveries. Materials which are to be discharged in the open on-site, i.e. aggregates, should be accompanied by someone in authority until actually unloaded from the delivery vehicles.

Plant control

1. Plant Utilisation Sheets – are sometimes required by head office to check on the usage of all plant on-site. Active time, idle time, breakdown time and maintenance time should be recorded.

   To reduce costs, site managers should be encouraged to return plant which remains idle, and discouraged from hoarding on the off-chance that the plant may be required later.
2. Maintenance – this falls into the following categories:
   (*a*) Servicing – cleaning, oil and water checks, and the odd adjustments to screws, bolts, etc.
   (*b*) Preventative maintenance – this routine maintenance is undertaken by a fitter/mechanic who will generally inspect the exposed parts of a machine and replace oil filters and plugs, and parts which are known to wear out at certain intervals.
   (*c*) Planned maintenance – when a machine ceases to work efficiently, arrangements will be made to give it a major overhaul.
3. Usage – where a tower crane is involved, a system of booking its use should be available and would be the site manager's responsibility to ensure maximum and equal usage by each trade. Where an activity is critical, priority should be given to it.
4. Plant Transfer Sheets – copy of this form sent to head office and to site where plant is being transferred (see Fig. 11.1.3).

Subcontractors' control

If a site is well organised, arrangements would be made to ensure continuity of work for the subcontractors. Where there is an anticipated delay which will

BUILDING CONTRACTORS LTD. MERROW, GUILDFORD 2657

Transfer of plant

No:482

Date:

From site:	To site:

Description	Quantity

Issued by:	Received by:	Drivers signature:

This document is made out in triplicate; the top copy sent to head office; second copy to the site which is to receive the plant; third copy retained

Fig. 11.1.3

affect the subcontractors, adequate early warning will assist them to make alternative arrangements to deploy their labours elsewhere. Advanced notice should be given to conform with earlier agreements of their expected arrival on-site. Adequate facilities must exist for the subcontractors' representatives to meet and discuss any problems which inevitably occur; this should be by a direct approach to the site manager and at site meetings.

Details about each subcontractor should be given to the contracts manager and site manager as shown in Fig. 11.1.8.

Personnel control

1. Timekeeping – this can be organised by one of a number of ways, either by the operatives clocking-on, signing-on or merely shouting their works number through the timekeeper's office window.

   Where regular lateness of an operative occurs, an adequate level of warning should be given in accordance with the Working Rules Agreement.
2. Personal records – records may be kept on-site regarding each operative on such matters as: extra payments allowed, warnings for misconduct, absences from work and any other relative information which will assist management to maintain control, particularly in cases of grievances, disputes or claims against the company by the operatives (see Fig. 11.1.4).
3. Bonus – realistic targets should be agreed with the operatives or their union representatives to enable a fair return for their conscientious labours. Additionally, this would encourage trade foremen and gangers to arrange, in advance, for adequate levels of materials to be on hand for

their gangs, and to plan for smooth transfers from one operation to another to help maintain suitable bonus earnings.

Many disputes occur on-site because the operatives become frustrated in trying to maintain their level of earnings, caused by mismanagement.

4. Non-productive time – during inclement weather or low temperatures, the site manager determines when work should temporarily cease.

Progress control

A programme is prepared as an overall illustration of the work to be undertaken on a contract. If the operations in the programme are broken down into finer details, it can be used more effectively as a control document. In breaking down each operation into further activities, additional programmes to the outline or contract programme can be drawn up by the site manager assisted by the programming officer. These programmes are known as:

1. Stage programmes – at various stages during the contract work (each stage being six weeks), the operations are broken down into a number of activities to show a more detailed description of the work to be done during each stage. Adjustments are best made in the stage programmes, particularly where work is behind programme.
2. Weekly programmes – stage programmes can be reduced even further to show weekly proposed activities. The weekly programmes are presented in chart form, separate ones being prepared for each trade, and are issued to the foremen/gangers. An alternative to preparing charts would be to give written instructions each Friday to the supervisors which would account for the following week's work for each trade.
3. Daily programmes – written (duplicated) instructions are prepared and issued to each foreman/ganger by the site manager; as an alternative, verbal instructions can be given.

Recording progress

There are various techniques used by contractors to show the level of progress of a project. One method is to colour in the percentage completed spaces (shown beside the operations in Fig. 10.3.1) each week. Another method is to colour in the duration bars on the programme, provided that the bars were not shaded in when the programme was prepared. If a vertical cursor (coloured string) were moved along the contract programme at the end of each week, it would help to highlight those operations in advance but particularly those in arrears.

Meetings

1. Domestic site meetings could be held weekly, especially where the site work is complicated or the contract is a large one, to discuss ways of maintaining progress. If certain operations are falling behind schedule, methods of rectification can be formulated and agreed. Where shortages

Side 1

Building Contractors Ltd.

OPERATIVE'S PERSONAL RECORD CARD

Surname: ____________ Forenames: ____________

Home Address: ____________

____

Date of Birth: ____________ N.I. No.: ____________

Works number: ____________ Trace: ____________

Date of engagement: ____________ Recruitment point: ____________

Other information: (Lodging address, allowances, expenses, etc.)

____

____

____

Warnings

Date	Type	Reason	Sign	Date	Type	Reason	Sign

Termination of employment

Date: ____________ Site: ____________

Reason for leaving: ____________

Assessment: ____________

Signed: ____________

Side 2

Transfers and rate increases

Date	Site	Occupation	Inc. Rate	Remarks	Sign

Sickness and other absences

From	To	From	To	From	To

Periodic travel

Date	Date	Date	Date	Date	Date

Courses

Date	Title	Date	Title	Date	Title

Fig. 11.1.4

of labour, plant, materials and information have developed, these problems can be discussed and action taken to rectify them. Lastly, the following week's programmes can be decided and instructions issued by the site manager. Those present at the meeting, which is called by the site manager, usually consists of the programming officer, general foremen, and if necessary, the trades foremen and gangers.

2. Site meetings are arranged by the contracts' manager when necessary to review progress with representatives from the suppliers, subcontractors and others associated with the project. If necessary, the architect or his representatives can be invited to attend.

   The programme can be discussed, particularly where there are material delivery date problems, or where a subcontractor has found difficulty in maintaining a suitable flow of work.

   If urgent information is required by any of those attending the meeting, notes should be taken to enable a follow up of requirements from the architect.

3. Project meetings (sometimes held on a monthly basis) are called by the architect who would expect the following to attend: professional quantity surveyor, consultants, clerk of works, contractor's representative, and, if necessary, major subcontractor's and supplier's representatives. All queries regarding the following would be discussed:
   (*a*) drawings, bills of quantities and specifications;
   (*b*) progress;
   (*c*) extension of time;
   (*d*) payments;
   (*e*) contractual claims – dayworks, variations, fluctuations, etc.;
   (*f*) quality of main contractor's work;
   (*g*) quality of subcontractors' work;
   (*h*) any other queries or problems to maintain good working relationships between all parties involved in the contract.

Other information

To ensure that the contractor's head office is kept informed of the progress of the site works and to help maintain a satisfactory level of Cost Control, various other documents are sent either on a daily, weekly, or monthly basis from site, such as:

1. Copy of the site Diary entries (duplicate) (see Fig. 11.1.5).
2. Weekly Report (if more detail is required than is shown in the site diary duplicate).
3. Copy of any formal site meeting minutes.
4. Time sheets.
5. Labour returns and increased costs sheet (see Fig. 11.1.6).
6. Copies of architect's instructions.
7. Copies of daywork sheets (see Fig. 11.1.7).
8. Copies of variation orders.
9. Bonus sheets.
10. Valuations of work completed.

3. Builder Ltd., Merrow Guildford — SITE DIARY

CONTRACT ______ No: ______ DATE ______

HOURS LOST ______ WEATHER ______ Temp. Max. ______ Min. ______

DIARY ______

VISITORS: ______

DRAWINGS RECEIVED ______

VARIATIONS/INSTRUCTIONS ______

LABOUR ON SITE

Own Labourers		Sub-Contractors	No.	Time	Sub-Contractors	No.	Time
Own Labourers		Bricklayers			Plasterers		
,, Bricklayers		Carpenters			Wall Tilers		
,, Carpenters		Roofers			Heating Engineers		
,, Fitters		Asphalters			Plumbers		
,, Painters		Steel Fixers			Electricians		
,, Plumbers		Painters			Glaziers		
		Floor Tilers					

ACTION REQUIRED BY BUYER :— (Or materials to be cleared)	PLANT REQUIRED :— (Or to be cleared)

Has anything happened, or not happened, today which has caused, will or may cause, delay or uneconomic working?
If so, please provide details. If not, say so.

______

Signed .... Site Manager.

Fig. 11.1.5

BUILDING CONTRACTORS LTD., MERROW, GUILDFORD

Labour returns and increased costs

Contract:

Contract No:

No:

Week ending:

For site use									For office use												
Name		M	T	W	T	F	Sat	Sun	Total		Rate		Guaranteed min. bonus		Holidays with pay		Earnings related contributions				Total
									Days	Hours	Increase	Amount	Increase	Amount	Increase	Amount	Increase	Amount			

This form can take account of any increased costs relating to labour on site if new agreements are made by employers with the operatives unions through the National Joint Council for the Building Industry.

Fig. 11.1.6

R. Builder Ltd Merrow, Guildford DAYWORK SHEET

Distribution: White – Architect; Yellow – Qty. Surveyor; Blue – Wickens Surveyor; Green – Wickens Site; Red – Main File

Contract ______________ Date __________ A.I. No. ______ SITE V.O. No. ______

DESCRIPTION OF WORK ____

LABOUR NAME	TRADE	M	T	W	T	F	S	S	TOTAL	RATE	£	p
										Total LABOUR To Summary £		

MATERIAL DESCRIPTION	QTY	RATE	£	p
		Total MATERIAL To Summary £		

PLANT (Including Transport) EQUIPMENT	M	T	W	T	F	S	S	TOTAL	RATE	£	p
									Total PLANT To Summary £		

SUMMARY	BROUGHT FORWARD	% ON COST	TOTALS £	p
LABOUR £				
PLANT £				
MATERIAL £				
TOTAL OF DAYWORK SHEET £				

Signed ______________________ Site Manager

Signed ______________________ Architect/C. of W.

Fig. 11.1.7

B.BUILDER & SON LTD

INTERNAL MEMO FROM: M.GARDNER TO ....SITE MANAGER
SUB-CONTRACT PURCHASER COPY ....CONTRACTS MANAGER
FILE

DETAILS OF SUB-CONTRACT CONDITIONS

1. SITE

2. NATURE OF WORK

3. NAME, ADDRESS & PHONE NO. OF SUB-CONTRACTOR

4. PROGRAMME DETAILS

(a) Period for executing work.

Item	Period

(b) Period of notice for commencement .... weeks.
(Note CM to confirm commencement date in writing).

(c) Approximate commencement date.

5. SUB-CONTRACT DETAILS

(a) Status of sub-contractor - Nominated/Domestic/Labour Only.

(b) Value of sub-contract work - £ Daywork rate - £

(c) Extent of work including list of relevant documents/bill pages.

(d) General attendance required.

(e) Special attendance required.

(f) Other facilities and conditions

NB. No other attendances and facilities will be provided by us.

Signed .... Date ....

Fig. 11.1.8

11. Subcontractors' claims (work done, including variations). Site manager should first check details of subcontractor conditions issued by the subcontract purchaser – see Fig. 11.1.8.
12. Any other documents which maintain an acceptable level of feed-back to enable actual output to be compared with the output forecast, adjustments being made for future estimates in the light of new experiences.

Safety, health and welfare controls

To ensure adequate controls within these areas, the site manager's main responsibilities are divided into:

1. Inspections.
2. Records.
3. Notifications.

See Chapter 7 for more detailed descriptions.

11.2 Operatives' employment conditions

It is recognised that certain principles should be observed by employers regarding operatives' conditions of employment, but unfortunately any agreements made between the operatives' unions and employers' union only takes effect where the operatives and employers alike belong to the organisations which are parties to the agreement laid down in the documents known as the Working Rule Agreements. However, when one considers the number of construction employers in existence in the United Kingdom and compares them with the number belonging to the BEC and FCEC and the many smaller employers' unions, there is still a large proportion who find it unnecessary to belong to such bodies (fortunately, this mainly applies to some of the very small firms). On the other hand, if one considers that generally there are over one million operatives, certainly not more than half belong to a trade union. With this apathy by some employers and employees, probably caused by the inability to take on further financial burdens (union subscription), some sections of the industry still maintain excellent working relationships because of standing agreements or working rules, while others progress at a slower rate, particularly regarding conditions for operatives.

Many employers obtain satisfactory results from their operatives by offering better conditions than those laid down by such bodies as the NJCBI in their National Working Rules. Unfortunately, other employers exist by paying less than the nationally agreed figures. This applies in some rural areas or within firms where operatives are lacking in union membership.

Employment conditions

Whether unions dominate or not, consideration should always be given by an employer and his/her site manager to the following areas which affect the employment conditions of operatives:

1. Conditions of service

The Employment Protection (Consolidation) Act 1978 and the Employment Protection Act 1975 (amended 1979) form the basis of minimum requirements when taking on new employees. A Contract of Employment need not be made in writing but will exist so long as an employee commences work and accepts the employer's conditions. There is, however, a requirement under the aforementioned Acts that a written Statement should be issued to an employee within thirteen weeks of commencement of work outlining his/her terms and conditions of employment, namely: the names of the parties, date of employment, job title, pay, hours of work, holidays, holiday pay, sickness benefits, pensions and pension schemes (if any), notices to terminate employment, and whether previous employment counts towards continuous service or not. In addition, the Statement should state any disciplinary rules or regulations requirements that exist within the firm, and the procedures to be adopted for airing grievances.

2. Periods of notice

Under the Employment Protection (Consolidation) Act 1978 and the Employment Protection Act 1975, the following notices are expected from the employer to terminate an employee's employment:

(*a*) minimum of 1 week where an employee has been employed continuously for 4 weeks or more;
(*b*) minimum of 2 weeks where an employee has been employed continuously for 2 years or more;
(*c*) minimum of 1 additional week's notice for every subsequent year's service up to 12 years (maximum notice being 12 weeks).

These conditions apply to anyone who is employed for 16 hours or more per week (subject to certain types of employment but which does not affect the construction industry).

3. Hours of work

In the construction industry, a 39-hour week is recognised as the maximum to be worked by operatives on normal rates of pay from Monday to Thursday at 8 hours per day, and Friday for 7 hours. Payments are not made for rest breaks during the day.

4. Earnings

Some firms pay basic wage rates to operatives with craftsmen being paid slightly more than skilled and unskilled workers. To help achieve better outputs, bonus schemes can be introduced which also enable operatives to earn more if they wish. A target rate for a job would be fixed (preferably by agreement), and if an operative completes the job in less time, any savings would be shared by the operative and employer in the proportions of, for example, 50/50 or 75/25.

Where a firm does not operate a bonus system which is 'geared' to production, then it is recommended that a 'standing' bonus be allowed.

5. *Overtime*

By agreement with the operatives, an employer may introduce overtime working to complete work earlier. In the majority of areas in the United Kingdom, overtime hourly rates are established which are higher than normal rates. If an operative exceeds a full day's work (8 hours), overtime payments would become due usually at the following rates:

(*a*) first 3 hours of overtime at 1½ times the normal hourly rate;
(*b*) any subsequent hours of overtime until starting time the next day at double the normal hourly rate.

Special established overtime rates would be paid for Saturday, Sunday and Bank Holiday working and numerous firms honour these rates. Unfortunately, in a few rural areas and within some firms which do not recognise union agreements, little attempt is made to pay for overtime at overtime rates.

6. *Guaranteed earnings*

With the unpredictable weather in certain periods during the year, the operatives sometimes find themselves working a few hours only each day. Many years ago, employers only paid for the actual hours worked; now, it is an acceptable practice, brought about by union pressure, to pay a full week's basic wages to operatives who make themselves available for work each day, and to this wage would be added the guaranteed bonus (standing bonus).

7. *Additional payments*

There are many varied types of work and working conditions within the industry. Construction operatives, like any other industrial worker, prefer not to work in conditions which are unhealthy, dangerous or inconvenient. When they are called upon to work in such conditions, extra payments are usually allowed by employers (where agreements exist between employers and unions, payments tend to be almost automatic). Sometimes, however, employers use their own discretion without consulting the operatives, which results in no payments being made.

Craftsmen require tools to work with and invariably provide their own. As the tools are expensive and cannot last for ever, employers, in the main, pay between 75p and £1.50 per week to each operative for what could be described as a 'hire charge' for their tools.

8. *Holidays*

These are divided into:

(*a*) annual holidays;
(*b*) public holidays.

Where agreements exist between the operatives' unions and employers' unions, a minimum of 21 days annual holiday and 8 days public holiday are normally allowed to each operative (2 weeks at Christmas, 1 week at Easter and 2 weeks some time in the summer). One would hope that employers outside the agreements afford their operatives similar holidays.

Holidays are paid for by the employer who either:

(*a*) set aside sums of money each week to be paid to the operatives at holiday periods, or
(*b*) holiday stamps are purchased from the Building and Civil Engineering Holidays Scheme Management Ltd, which are then affixed to each operative's holiday card (a surcharge paid by the employer, covers a death benefit scheme and allows a lump sum pension for the employees).

In the latter case, when an operative leaves his/her employment to take up another appointment, his/her holiday card (stamped up-to-date) is then transferred to the new employer, who continues to affix stamps each week. The stamps are traded in for cash by the operative when holidays are taken, and the face value is paid by the non-profit making holiday Stamp Company.

9. Travelling

Fares and travelling time payments may be paid by an employer when an operative is sent out to a job, but it usually depends on what distance the job is from the home of the operative. The payments are normally allowed for one way only, but some firms are generous and allow fares and travelling time both ways (to and from work). Naturally, if the firm laid on free transport, the only payment that an operative could expect is a travelling time allowance.

10. Misconduct

It is essential for site managers and Industrial Relations Officers to observe the firm's Handbook (if provided with one) when dealing with cases of misconduct by operatives, or site staff, particularly now that there is an ever increasing number of applications to the Industrial Tribunals claiming wrongful dismissal, redundancy and breaking of a contract.

The following is the normal procedure where there has been a breach of conduct:

(*a*) verbal warning in front of witnesses;
(*b*) written warning if misconduct continues;
(*c*) dismissal if misconduct persists – stating in writing the reason for dismissal.

Note: It must be remembered that this dismissal document could be used as evidence by the operative at the Industrial Tribunal, so reason for dismissal should be valid.

11.3 Joint consultations

During the running of a contract good industrial relations are of paramount importance to minimise or prevent disputes occurring between the firm and the operatives' trade unions. To this end, the National Joint Council for the Building Industry (NJCBI); the Civil Engineering Construction Conciliation Board (CECCB); and other similar bodies were created and are comprised of representatives from the operatives' and employers' unions.

The NJCBI is the largest representative body and the following organisations are members (see Fig. 2.4.1).

1. BEC (Building Employers Confederation).
2. UCATT (Union of Construction and Allied Trades and Technicians).
3. TGWU (Transport and General Workers Union).
4. GMBATU (General and Municipal, Boilermakers and Allied Trades Union).
5. FTATU (Furniture, Timber and Allied Trades Union).

The NJCBI functions to fix wages and rates of pay, and lays down the conditions of employment of operatives. Special committees, etc. have been established to the mutual benefit of both sides of the Council, which are:

1. Holidays with Pay Joint Committee.
2. National Joint Training Commission.
3. Conciliation Panels.

Many disputes and problems have been resolved by the Joint Committee before they have had time to develop into affairs which are unsatisfactory at the shop floor or site level. Representatives from both sides at this level, the site manager and union stewards, should have suitable training in industrial relations to make them aware of the importance of observing the Working Rule Agreement, which was prepared by the NJCBI and which is updated regularly as new legislation is introduced by the Government, and where new agreements are made between the operatives' and employers' representatives.

At site level, all the Working Rules are important, but the grievances procedures need to be stressed to all parties to prevent stoppages, walk-outs and work to rules during disputes which can be extremely costly to both sides and damaging to the Industry's image.

After the setting up of the Court of Enquiry during the well-known Barbican dispute in London (mid-1960s), a 'Report' was published which, apart from criticising the action of both the employers and operatives, recommended that employers should appoint 'trouble shooters' (Industrial Relations Officers) who could take positive action when allegations are made of mismanagement of employees or incorrect procedures are being adopted by operatives or union stewards. By the appointment of such officers the unions would then know with whom they have to deal, thereby satisfactory channels of communication are ensured and prompt action can then be taken by either side when grievances, disputes or differences occur.

Industrial Relations Officers should understand the principles behind the Working Rule Agreements and should be knowledgeable in the interpretation of the Rules. Site Managers must know the contents, particularly the Rules which affect working conditions on site, i.e. working hours; safety recommendations; disputes procedures; health and welfare facilities; trade union recognition and procedures, to mention but a few.

Labour relations policy

Each firm is advised to draw up a Labour Relations Policy for all managers to observe. Handbooks on Labour Relations Procedures can be prepared to distribute to supervisors for reference if and when problems arise in the works or on-site.

Because of the complexity of dealing with operatives on-site particularly due to the WRA and protective legislation covering their employment, Personal Record Cards should be kept, preferably on-site, for each individual so that records on the dates of engagement, transfers, warnings for misconduct and other matters can be made so that in the future correct procedures are maintained regarding:

1. Periods of notice – when terminating someone's employment recognised periods of notice depend on length of service.
2. Dismissals – previous dates of warning for misconduct can be checked.
3. Payment of wages – extra payments for responsibility, skill, lodgings and travelling can be made after referring to Records (see Fig. 11.1.4 for example of Personal Record Card).

Wants, 194
Waste calculations, 186
Windmills, 4
Work Study, 74
Working drawings, 102, 107
Working Rule Agreements, 84, 156, 253
World, map of, 12–13
Wren, Christopher, 36

Index